SpringerBriefs in Computer Science

Series editors

Stan Zdonik, Brown University, Providence, Rhode Island, USA
Shashi Shekhar, University of Minnesota, Minneapolis, Minnesota, USA
Xindong Wu, University of Vermont, Burlington, Vermont, USA
Lakhmi C. Jain, University of South Australia, Adelaide, South Australia, Australia
David Padua, University of Illinois Urbana-Champaign, Urbana, Illinois, USA
Xuemin (Sherman) Shen, University of Waterloo, Waterloo, Ontario, Canada
Borko Furht, Florida Atlantic University, Boca Raton, Florida, USA
V.S. Subrahmanian, University of Maryland, College Park, Maryland, USA
Martial Hebert, Carnegie Mellon University, Pittsburgh, Pennsylvania, USA
Katsushi Ikeuchi, University of Tokyo, Tokyo, Japan
Bruno Siciliano, Università di Napoli Federico II, Napoli, Italy
Sushil Jajodia, George Mason University, Fairfax, Virginia, USA
Newton Lee, Newton Lee Laboratories, LLC, Tujunga, California, USA

More information about this series at http://www.springer.com/series/10028

Fabio Fassetti · Simona E. Rombo
Cristina Serrao

Discriminative Pattern Discovery on Biological Networks

 Springer

Fabio Fassetti 🄳
University of Calabria
Cosenza
Italy

Cristina Serrao
University of Calabria
Cosenza
Italy

Simona E. Rombo
University of Palermo
Palermo
Italy

ISSN 2191-5768   ISSN 2191-5776   (electronic)
SpringerBriefs in Computer Science
ISBN 978-3-319-63476-0   ISBN 978-3-319-63477-7   (eBook)
DOI 10.1007/978-3-319-63477-7

Library of Congress Control Number: 2017947457

Printed on acid-free paper

This Springer imprint is published by Springer Nature
The registered company is Springer International Publishing AG
The registered company address is: Gewerbestrasse 11, 6330 Cham, Switzerland

Preface

Biological networks are an effective model for providing insights about biological mechanisms. Networks with different characteristics are employed for representing different scenarios. This powerful model allows analysts to perform many kinds of analyses which are proved to mine interesting information about underlying biological behaviors. This work intends to survey on biological networks as a model for subsequent analyses which are also presented and discussed. Then, it focuses on techniques for discovering exceptional patterns, a particular kind of analysis which is witnessing a great interest due to the importance of the knowledge it is able to mine, with a pattern accounting for local similarities and also collaborative effects involving interactions between multiple actors (for example, genes). Among exceptional patterns, of particular interest are discriminative ones, namely patterns which are able to discriminate between two input populations (for example, healthy/unhealthy samples). The work will also include discussions on the most recent proposal on discovering discriminative patterns, where there is a labeled network for each sample, resulting in a database of networks representing a sample set. Previous techniques are able to just consider an aggregated network of each population, thus allowing the analyst to get a much more fine analysis. In more details, edge-labeled networks are used and the discriminative power of a pattern is measured based on edge weights, which are representative of how much relevant is the co-expression between two genes.

Rende, Italy
Palermo, Italy
April 2017

Fabio Fassetti
Simona E. Rombo
Cristina Serrao

Acknowledgements

This research has been partially supported by the PRIN project 20122F87B2 titled "Compositional Approaches for the Characterization and Mining of Omics Data" co-financed by the Italian Ministry of Education, University and Research and INdAM—GNCS Project 2016: "Approcci integrativi e computazionali per lestrazione di conoscenza da reti funzionali".

Contents

Part I
Biological Networks

Chapter 1
Data Sources and Models

Abstract Biological networks rely on the storage and retrieval of data associated to the physical interactions and/or functional relationships among different actors. In particular, the attention may be on the interactions among cellular components, such as proteins, genes, RNA, or for example on phenotype–genotype associations. Data from which biological networks are built are usually stored in public databases, and we provide here a brief summary of the main types of both data and associations, publicly available. Moreover, we also explain how it is possible to construct suitable network models from these associations, focusing on protein–protein interaction networks, gene–disease networks and network populations. Being the latter the model we adopted for the Part II of this book, we describe more details about it.

Keywords Biological datasets · Interaction databases · Protein–protein interactions · Gene co-expression · Biological data · Interaction data

1.1 Protein Data

Protein–protein interactions are physical interactions between pairs of proteins detected by many different experimental approaches, also including high-throughput experiments [10, 11]. This type of interaction is usually available in "mitab" format, that is, a tab-separated file where each column contains information about the interactors, the experiment from which the interaction has been detected, a possible reliability value scoring how much reliable is that interaction based on independent information, and other functional annotations. Among the most important protein–protein interaction sources we mention INTACT [5], BIOGRID [1], and HPRD [6].

1.1.1 Protein–Protein Interaction Networks

The set of all the protein–protein interactions of a given organism is its *interactome*, usually modeled by an indirect graph, called *protein–protein interaction network* (PPI

F. Fassetti et al., *Discriminative Pattern Discovery on Biological Networks*,
SpringerBriefs in Computer Science, DOI 10.1007/978-3-319-63477-7_1

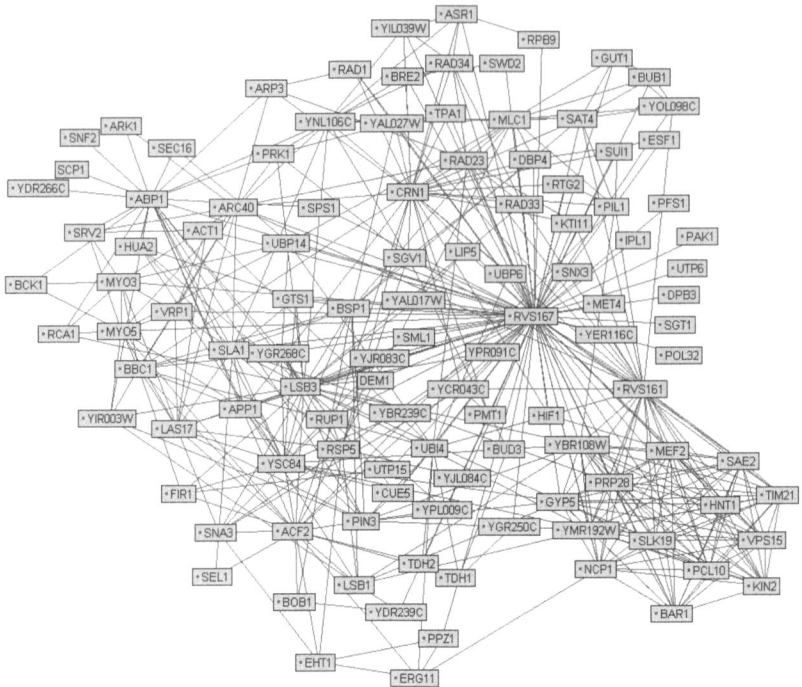

Fig. 1.1 A small portion of the *Saccharomyces cerevisiae* interactome, drawn by using PIVOT Software [9]

network), where nodes represent involved proteins and edges encode their interactions (see Fig. 1.1).

Nodes can be labeled by protein names or ids, while suitable reliability scores can be used as edge labels. Indeed, many databanks provide interaction reliability scores obtained by combining information coming from both the confidence of the techniques applied to discover a specific interaction and the fact that the same interaction is confirmed by different experimental techniques.

In many applications, PPI networks are also represented by their corresponding adjacency matrices.

1.2   Disease Data

Identifying genes involved in the onset of a disorder is the first step to understand the diseases mechanisms. Meta-analysis of published genetic associations, together with the new genome-wide association studies, has provided a huge amount of information on "risk alleles" and on genetic associations between genes and diseases that is cataloged in the OMIM database [4]. A single gene can influence many pathologies,

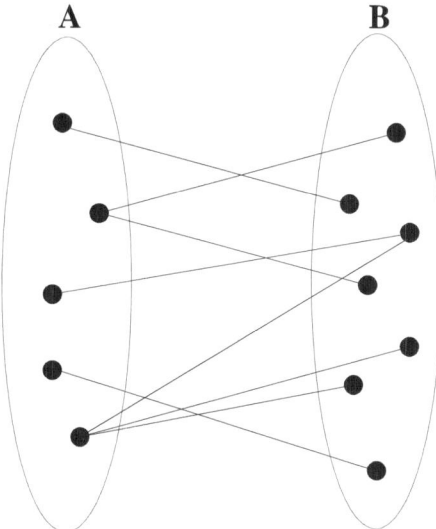

Fig. 1.2 A bipartite graph

and at the same time human diseases are often the consequence of the perturbation of multiple cellular components. Moreover, when two or more genes are associated to the same disorder, the corresponding proteins show a high propensity to interact [3].

1.2.1 Genotype–Phenotype Networks

Disease networks may be represented by a bipartite graph (see Fig. 1.2), built on the genetic diseases and the disease genes [3].

From this bipartite graph, two networks may be obtained: the first one with the diseases as nodes while edges linking two nodes indicate that they have in common at least one gene, and the second one by using the genes as nodes and linking two genes when associated to the same disorder. According to other representations [7], a metabolic disease network may be built, where disorders are linked if the corresponding mutated enzymes are involved in related pathways.

1.3 Co-expression Data

Another important source of data is co-expression data. A network can be built as follows: given a set of genes and a set of samples, where each sample has an expression value associated with the gene at hand, there is a node for each gene and an arc between two genes g_1 and g_2 if the values scored by the samples on g_1 and the values scored by the samples on g_2 are correlated. Note that the resulting network is a single network that represents the whole population.

In order to build this kind of networks an interesting data source is GEO [8], a public functional genomics data repository where microarray, next-generation sequencing, and other forms of high-throughput functional genomics data submitted by the research community are avaliable and downloadable for free.

GEO provides Sample Records and Series Records. The former describe the characteristics of each single sample analized, the latter link together a group of related samples and provides a description of the whole study. Sample records are sometimes joined in DataSet Records that represent a curated collection of biologically and statistically comparable samples and can be studied through many analysis tools provided by GEO.

Data retrievable in GEO repository can be used to compare sets of samples in order to identify genes that are differentially expressed.

1.3.1   Network Populations

An approach related to that previously described but substantially different is that proposed in [2]. Given a set of genes and a set of samples, where each sample has an expression value associated with the gene at hand, authors of [2] propose a method to build an independent network for each sample instead of collapsing the information to build a unique network representative of the population. In this case, in order to build the network for a single sample s, one has to decide if there is an arc between two genes g_1 and g_2 on the basis of just two values (the expression values of the sample s on g_1 and g_2). Therefore, the correlation or any other statistical measure cannot be directly applied. Thus, the proposal technique tries to measure whether the two values are in agreement or not and if this agreement is strong and statistically relevant. More details about the technique are provided in Chap. 4.

References

1. ChatrAryamontri, A., Breitkreutz, B.J., Heinicke, S., et al.: The BioGRID interaction database: 2013 update. Nucl. Acids Res. **41**(Database issue), D816–D823 (2013)
2. Fassetti, F., Rombo, S.E., Serrao, C.: Discovering discriminative graph patterns from gene expression data. In: SAC 2016, pp. 23–30. ACM (2016)
3. Goh, K.I., Cusick, M.E., Valle, D., Childs, B., Vidal, M., Barabasi, A.L.: The human disease network. Proc. Natl. Acad. Sci. USA **104**(21), 8685–8690 (2007)
4. Hamosh, A., Scott, A.F., Amberger, J., Bocchini, C., Valle, D., McKusick, V.A.: Online mendelian inheritance in man (omim), a knowledgebase of human genes and genetic disorders. Nucl. Acids Res. **30**(1), 52–55 (2002)

5. Kerrien, S., Aranda, B., Breuza, L., et al.: The IntAct molecular interaction database. Nucl. Acids Res. **40**(Database issue), D841–D846 (2012)
6. Keshava Prasad, T.S., Goel, R., Kandasamy, K., et al.: Human protein reference database–2009 update. Nucl. Acids Res. **37**(Database issue), D767–D772 (2009)
7. Lee, D.S., Park, J., Kay, K.A., Christakis, N.A., Oltvai, Z.N., Barabsi, A.L.: The implications of human metabolic network topology for disease comorbidity. Proc. Natl. Acad. Sci. **105**(29), 9880–9885 (2008)
8. NCBI: Gene Expression Omnibus. https://www.ncbi.nlm.nih.gov/geo/
9. Orlev, N., Shamir, R., Shiloh, Y.: PIVOT: protein interacions visualization tool. Bioinformatics **20**(3), 424–425 (2004)
10. Rigaut, G., Shevchenko, A., Rutz, B., Wilm, M., Mann, M., Seraphin, B.: A generic protein purification method for protein complex characterization and proteome exploration. Nat. Biotechnol. **17**(10), 1030–1032 (1999)
11. Walhout, A.J., Boulton, S., Vidal, M.: Yeast two-hybrid systems and protein interaction mapping projects for yeast and worm. Yeast **17**(2), 88–94 (2000)

Chapter 2
Problems and Techniques

Abstract When biological networks are considered, the extraction of interesting knowledge often involves subgraphs isomorphism check that is known to be NP-complete. For this reason, many approaches try to simplify the problem under consideration by considering structures simpler than graphs, such as trees or paths. Furthermore, the number of existing approximate techniques is notably greater than the number of exact methods. In this chapter, we provide an overview of three important problems defined on biological networks: network alignment, network clustering, and motifs extraction from biological networks. For each of these problems, we also describe some of the most important techniques proposed to approach them.

Keywords Biological network analysis · Graph alignment · Protein–protein interaction network clustering · Community search · Graph motif extraction · Global and local alignment

2.1 Network Alignment

Let N_1 and N_2 be two input networks. The alignment problem consists of finding a set of conserved edges across N_1 and N_2, leading to a (non-necessarily connected) conserved subgraph between them. In this case, the problem is also known as *pairwise alignment*. *Multiple alignments* is an extension of pairwise alignment such that a set of networks $N_1, \ldots, N_n$ is considered in input, and it is usually computationally more difficult to perform. In the following we refer to pairwise network alignment, and all the notions we will introduce can extend to multiple alignment.

The problem of biological network alignment can be distinguished in *global alignment* and *local alignment*. Global alignment aims at finding a unique (possibly, the best one) overall alignment between N_1 and N_2, in such a way that a one-to-one correspondence is found between nodes in N_1 and nodes in N_2. The result is made of a set of pairs of non-overlapping subgraphs of N_1 and N_2. Local alignment aims instead at finding multiple, unrelated regions of isomorphism among the input networks, each region implying a mapping independently of the others. Therefore,

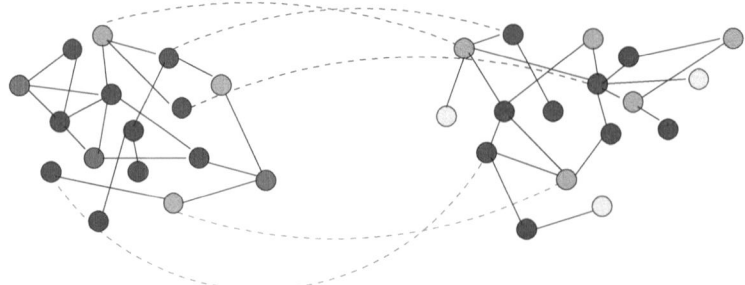

Fig. 2.1   Global alignment

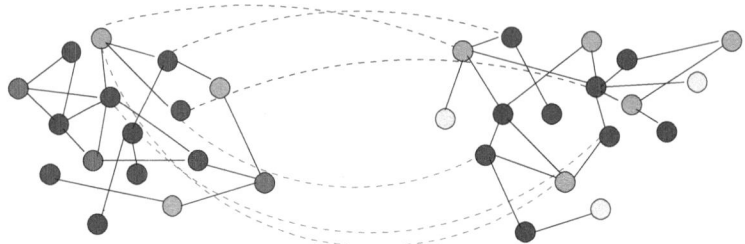

Fig. 2.2   Local alignment

the found correspondences may involve overlapping subgraphs. Figures 2.1 and 2.2 illustrate global alignment and local alignment, respectively.

Network alignment can be approached also if N_1 and N_2 are of different types, leading to a kind of *heterogeneous* alignment. Usually, in this case, the two input networks are merged and statistical approaches are then applied to extract the most significant subgraphs from the integrated network.

2.1.1   Techniques

Network alignment involves the problem of subgraph isomorphism checking, that is known to be NP-complete [27]. Therefore, the proposed techniques are often based on approximate and heuristic algorithms.

2.1.1.1   Global Alignment

Singh et al. [65] present *IsoRank*, an algorithm for pairwise global alignment of PPI networks working in two stages: it first associates a score with each possible match between nodes of the two networks and, then, constructs the mapping for the global network alignment by extracting mutually consistent matches according to a bipartite graph weighted matching performed on the two entire networks. *IsoRank* has been extended in [64] to perform multiple alignments by approximate multipartite graph weighted matching. In [39] the *IsoRankN* (IsoRank-Nibble) tool is proposed, that is, a global multiple-network alignment tool based on spectral clustering on the induced graph of pairwise alignment scores. In [34] a graph-based maximum structural matching formulation for pairwise global network alignment is introduced, combining a Lagrangian relaxation approach with a branch-and-bound method. MI-GRAAL [36] can integrate any number and type of similarity measures between network nodes (e.g., sequence similarity, functional similarity, etc.) and find a combination of similarity measures yielding the largest contiguous (i.e., connected) alignments. In [62] a scalable algorithm for multiple alignment is presented based on clustering methods and graph matching techniques to detect conserved interactions while simultaneously attempting to maximize the sequence similarity of nodes involved in the alignment. Finally, in [44] an evolutionary-based global alignment algorithm is proposed, while in [45] a greedy method is used, based on an alignment scoring matrix derived from both biological and topological information of the input networks to find the best global network alignment.

2.1.1.2 Local Alignment

Kelley et al. [31] proposed *PathBLAST*, that is, a procedure for pairwise alignment combining interaction topology and protein sequence similarity. They search for high scoring pathway alignments involving two paths, one for each network, in which proteins of the first path are paired with putative homologs occurring in the same order in the second path. *PathBLAST* is extended in [59] for multiple alignments, based on the generation of a network alignment graph where each node consists of a group of sequence-similar proteins, one for each species, and each link between a pair of nodes represents a conserved protein interaction between the corresponding protein groups. *PathBLAST* has also been used in [8] to resolve ambiguous functional orthology relationships in PPI networks. In [35], a technique for pairwise alignment is proposed based on duplication/divergence models and on efficient heuristics to solve a graph optimization problem. *Bi-GRAPPIN* [23] is based on maximum weight matching of bipartite graphs, resulting from comparing the adjacent nodes of pairs of proteins occurring in the input networks. The idea is that proteins belonging to different networks should be matched looking not only at their own sequence similarity, but also at the similarity of proteins they significantly interact with. In [24] an algorithm for multiple alignments, named *Graemlin*, is presented. *Graemlin* aligns an arbitrary number of networks to individuate conserved functional modules, greedily assigning the aligned proteins to non-overlapping homology classes and progressively aligning multiple input networks. The algorithm also allows searching for different conserved topologies defined by the user. In [14] the algorithm *C3Part-M*, based on a non-heuristic approach exploiting a correspondence multigraph formalism to extract connected components conserved in multiple networks, is presented and compared with *NetworkBlast-M* [30], another technique recently proposed based on a novel representation of multiple networks that is only linear in their size.

Finally, *AlignNemo* is proposed in [13] that builds a weighted alignment graph from the input networks, extracts all connected subgraphs of a given size from the alignment graph, and uses them as seeds for the alignment solution, by expanding each seed in an iterative fashion.

2.1.1.3 Other Approaches

ABiNet [19, 20] is an algorithm performing *asymmetric alignment*. In particular, given two input networks, the one associated to the best-characterized organism (called *Master*) is exploited as a fingerprint to guide the alignment process to the second input network (called *Slave*), so that generated results preferably retain the structural characteristics of the Master network. Technically, this is obtained by generating from the Master a finite automaton, called *alignment model*, which is then fed with (linearization of) the Slave for the purpose of extracting, via the Viterbi algorithm, matching subgraphs. *ABiNet* performs both querying and global alignment.

Finally, the approach [67] has been proposed in order to align heterogeneous networks, for example, PPI and disease networks.

2.1.2 Querying

Network querying consists on analyzing an input network, called *target network*, searching for the occurrences of a *query network* of interest. Such a problem "is aimed at transferring biological knowledge within and across species" [60], since the found subnetworks may correspond to cellular components involved in the same biological processes or performing similar functions than the components in the query.

2.1.2.1 Techniques

Network querying approaches may be divided into two main categories: those ones searching for efficient solutions under particular conditions, e.g., the query is not a general graph but it is a path or a tree, and other approaches where the query is a specific small graph in input, often representing a functional module of another well-characterized organism. *MetaPathwayHunter* [50] is an algorithm to query metabolic networks by multi-source trees that are directed acyclic graphs whose corresponding undirected graphs are trees where nodes may present both incoming and outgoing edges. In [16, 63] *QPath* and *QNet* are presented, respectively. *QPath* queries a PPI network by a query pathway consisting of a linear chain of interacting proteins belonging to another organism. The algorithm works similarly to sequence alignment, by aligning the query pathway to putative pathways in the target network, so that proteins in analogous positions have similar sequences. *QNet* is an extension of *QPath* in which the queries are trees or graphs with limited treewidth. In [68] the two problems of path matching and graph matching are considered. An exact algorithm called *SAGA* is presented to search for subgraphs of arbitrary structure in a large graph, grouping related vertices in the target network for each vertex in the query. *NetMatch* [21] is a Cytoscape plugin allowing for approximated queries, that is, graphs where some nodes are specified and others are wildcards (which can match an unspecified number of elements). *NetMatch* captures the topological similarity between the query and target graphs, without taking into account any information about node similarities. In [22] a technique is proposed based on maximum weight matching of bipartite graphs. *Torque* [10] is an algorithm based on dynamic programming and integer linear programming to search for a matching set of proteins that are sequence-similar to the query proteins, by relaxing the topology constraints of the query. Finally, note that, sometimes, methods for local alignment can be also successfully exploited to perform network querying, e.g., [20, 31, 35].

2.2 Network Clustering

The analysis of protein–protein interaction networks may result in the detection of protein complexes helping in understanding the mechanisms regulating cell life, in describing the evolutionary orthology signal (e.g., [29]), in predicting the biological functions of uncharacterized proteins, and, more importantly, for therapeutic purposes. The problem of detecting protein complexes using PPI networks can be computationally addressed by using clustering techniques aiming at grouping together proteins which share a large number of interactions. Possible uncharacterized proteins in a cluster may be assigned to the biological function recognized for that module, and groups of proteins performing the same tasks can be singled out this way.

PPI networks have various characteristics which have to be taken into account when developing clustering algorithms for detecting functional complexes.

2.2.1 Techniques

MCODE (*Molecular COmplex DEtection*) [7] relies on a node weighting procedure by local neighborhood density and outward traversal from a locally dense seed protein, in order to isolate the dense regions according to given input parameters. The algorithm allows fine-tuning of clusters of interest without considering the rest of the network and allows examination of cluster interconnectivity, which is relevant for protein networks. The algorithm may use a "fluff" option which increases the size of the modules to find and allows for overlapping among the output complexes, since the nodes added to a cluster are not marked as already used.

CFINDER [1] is based on the clique percolation concept (see [15, 46]). The idea behind this method is that a cluster can be interpreted as the union of small fully connected subgraphs that share nodes, where a parameter is used to specify the minimum number of shared nodes. CFinder extracts all the maximal complete subgraphs, i.e., the maximal cliques in the input PPI network. Then a clique–clique overlap matrix is built such that each entry contains the number of common nodes between the two corresponding cliques, and each diagonal entry is the clique size k. The k-cliques communities can be found by deleting every entry off the diagonal having a value less than $k - 1$, and every diagonal entry less than k. The remaining separate components will be the k-cliques communities. CFinder allows overlap between communities.

The greedy local expansion methods RANCoC [55], MF- PINCoC [52], and PINCoC [53] expand a single protein randomly selected by adding/removing connected proteins that best contribute to improve a given quality function based on the concept of co-clustering [40]. In order to escape poor local maxima, with a given probability, the protein causing the minimal decrease of the quality function is removed in MF- PINCoC [52] and PINCoC [53]. Instead RANCoC removes, with a fixed probability, a protein at random, even if the value of the quality function diminishes. This strategy is more efficient in terms of computation than that applied in the methods [52, 53], and it is more efficacious in avoiding entrapments in local optimal solutions. All the three algorithms work until either a preset of maximum number of iterations is reached, or the solution cannot further be improved. Both MF- PINCoC and RANCoC [55] allow overlapping clusters.

A typical instance of cost-based local search is RNSC (*Restricted Neighborhood Search Clustering*) [33], which explores the solution space of all the possible clusterings in order to minimize a cost function reflecting the number of inter-cluster and intra-cluster edges. The algorithm begins with a random clustering, and attempts to find a clustering with the best cost by repeatedly moving one node from a cluster to another one. A list of tabular moves is used to forbid cycling back to previously examined solutions. In order to output clusters likely to correspond to true protein complexes, thresholds for minimum cluster size, minimum density, and functional homogeneity must be set. Only clusters satisfying these criteria are given as the final result. This obviously implies that many proteins are not assigned to any cluster.

Several community discovery algorithms have been proposed based on the optimization of a modularity-based function (see e.g., [25]). Modularity measures the fraction of edges falling within communities, subtracted by what would be expected if the edges were randomly placed. In particular, QCUT [58] is an efficient heuristic algorithm applied to detect protein complexes. QCUT optimizes modularity by combining spectral graph partitioning and local search. By optimizing modularity, communities that are smaller than a certain scale or have relatively high inter-community density may be merged into a single cluster. In order to overcome this drawback, the authors introduce an algorithm that recursively applies QCUT to divide a community into subcommunities. In order to avoid over-partitioning, a statistical test is applied to determine whether a community indeed contains intrinsic subcommunity.

One of the first methods based on flow simulation for detecting protein complexes in a PPI network is the *Markov Clustering algorithm MCL* [17, 66]. MCL is based on the concept of a random walk on a graph to retrieve cluster structure and uses algebraic operations on a distance matrix associated with the graph. In a random walk the direction to be followed at each node is given by chance. MCL simulates many random walks (or flows) within a graph by strengthening flow where it is strong, and weakening it where it is weak. By repeating this process, a number of regions with strong internal flow (the clusters), separated by boundary with no flow, will appear. The flow is simulated by algebraic operations on a stochastic Markov matrix, such as flow expansion and an inflation operator which raises each entry of the matrix to a given power, and then rescales the matrix so that the column sum equals 1. By repeating a number of times squaring, inflating, and scaling, the matrix tends to an equilibrium state that shows the cluster structure. The inflation parameter influences the number of clusters.

In [51, 54] genetic algorithms have been applied to PPI networks, referred as *GA-PPI*, by performing an extensive experimental evaluation aiming at exploring the capability of genetic algorithms to find clusters in PPI networks, when different topological-based fitness functions are employed. The adopted representation of individuals is the graph-based adjacency representation, originally proposed in [49], where an individual of the population consists of n genes, each corresponding to a node of the graph modeling the PPI network. A value j assigned to the ith gene is interpreted as a link between the proteins i and j, and implies that i and j belong to the same cluster. In particular, in [51] the fitness functions of *conductance, expansion, cut ratio, normalized cut*, reported from [38], are employed, while in [54], the cost functions of the $RNSC$ algorithm [33] have been used.

2.3 Network Motif Extraction

The concept of *motif* has been exploited in different applications of computational biology [3, 47]. Depending on the context, *what is a motif* may assume sensibly different meanings. In general, motifs are always associated to repetitive objects. For example, a repeated substring can be considered a motif when its frequency is greater than a fixed threshold, or instead when it is much different than expected [5].

Also in the context of biological networks, a motif can be defined according to its *frequency* or to its *statistical significance* [12]. In the first case, a motif is a subgraph that appears more than a threshold number of times in an input network; in the second case, a motif is a subgraph that appears more often than expected by chance. In particular, to measure the statistical significance of the motifs, many works compare the number of appearances of the motifs in the biological network with the number of appearances in a number of randomized networks [18], by exploiting suitable statistical indices such as *p value* and *z score* [43].

Despite the similarity with sequences, that is evident from the two definitions above, network motifs present important differences w.r.t. string motifs, the main important of which concerns the computational complexity of the problem of motif extraction, that is, polynomial for strings and exponential (in the size of the input) for networks.

Given a biological network N, a *motif* can be defined according to its *frequency* or to its *statistical significance* [12]. In the first case, a motif is a subgraph appearing more than a threshold number of times in N; in the second case, it is a subgraph occurring more often than expected by chance. In particular, to measure the statistical significance of a motif, many works compare its number of occurrences with those detected in a number of randomized networks [18], by exploiting suitable statistical indices such as *p value* and *z score* [43].

2.3.1   Techniques

Shen-Orr et al. [61] defined *network motifs* as "patterns of interconnections that recur in many different parts of a network at frequencies much higher than those found in randomized networks." They discovered three highly significant motifs composed by three–four nodes among which the most famous is the "feed-forward loop," whose importance has been shown also in further studies [41, 42]. The technique presented in [61] laid the foundations for different extensions, such as [9, 11, 69]. In [69], composite motifs consisting of two kinds of interactions are extracted by exploiting edges of different colors in the network modeling. In particular, two types (colors) of edges are considered, representing protein–protein and transcription–regulation interactions, and algorithms are developed for detecting network motifs in networks with multiple types of edges. In [9], topological motifs derived from families of mutually similar, but not necessarily identical, patterns are discussed and extracted. The authors developed a search algorithm to extract topological motifs called *graph alignment*, in analogy to sequence alignment, that is based on a scoring function. All the approaches mentioned above relate the concept of *motif* only to the network topology. As observed in [37], there are biological networks (e.g., metabolic networks) where a purely topological definition of motifs seems to be inappropriate as similar topologies can give rise to very different functions. Thus, the authors of [37] introduce a new definition of motifs in the context of metabolic networks, such that the components of the network play the central part and the topology can be added only as a further constraint. Similarly to [37], in [48] the concept of motif is related to both graph structure and node similarity. In particular, the author presents a three-step exact approach based on the application of the notion of maximality, used extensively in strings and arrays [2, 4, 6, 26, 28, 56, 57], to graphs. In [32] the two notions of *structural* and *biological network motifs* are distinguished, focusing on the latter one that the authors explain as biologically significant small connected subgraphs regardless of the structure. They introduce five algorithms for the discovery of biological network motifs reducing the number of subgraphs to search by removing a number of edges from the original network and, at the same time, increasing the discovery rate for biological network motifs.

The search of significant motifs in biological networks is pioneered by Shen-Orr et al. [61], where *network motifs* have been defined as "patterns of interconnections that recur in many different parts of a network at frequencies much higher than those found in randomized networks." The authors of [61] studied the transcriptional regulation network of *Escherichia coli*, by searching for small motifs composed by three–four nodes. In particular, three highly significant motifs characterizing such network have been discovered; the most famous is the "feed-forward loop," whose importance has been shown also in further studies [41, 42].

The technique presented in [61] laid the foundations for different extensions, the main of which are [9, 11, 69].

In [69], composite motifs consisting of two kinds of interactions have been taken into account by exploiting edges of different colors in the network modeling. They modelled an integrated cellular interaction network by two types (colors) of edges, representing protein–protein and transcription–regulation interactions, and developed algorithms for detecting network motifs in networks with multiple types of edges.

In [9], topological motifs derived from families of mutually similar, but not necessarily identical, patterns have been discussed and extracted from the gene regulatory network of *Escherichia coli*. The authors developed a search algorithm to extract topological motifs called *graph alignment*, in analogy to sequence alignment that is based on a scoring function.

In [11], n-nodes "bridge" and "brick" motifs are searched for in complex networks, and the presence of such motifs has been associated with network topology, but not with network size. The authors proposed a method for performing simultaneously the detection of global statistical features and local connection structures, and the location of functionally and statistically significant network motifs.

All the approaches [9, 11, 61, 69] relate the concept of *motif* only to the network topology, without any consideration of possible properties shared by network nodes in terms of their mutual similarity.

As observed in [37], there are biological networks (e.g., metabolic networks) where a purely topological definition of motifs seems to be inappropriate as similar topologies can give rise to very different functions. Thus, the authors of [37] introduce a new definition of motifs in the context of metabolic networks, such that the components of the network play the central part and the topology can be added only as a further constraint.

Similarly to [37], in [48] the concept of motif is related to both graph structure and node similarity. In particular, the author presents a three-step exact approach based on the application of the notion of maximality, used extensively in strings, to graphs.

The two works [37, 48] open the way for the definition of new, exact or approximate, approaches for motif extraction taking into account not only the network topology but also the biological properties of the interacting components.

References

1. Adamcsek, B., et al.: CFinder: locating cliques and overlapping modules in biological networks. Bioinformatics **22**(8), 1021–1023 (2006)
2. Amelio, A., Apostolico, A., Rombo, S.E.: Image compression by 2D motif basis. In: Data Compression Conference (DCC'11), pp. 153–162 (2011)
3. Apostolico, A., et al.: Finding 3d motifs in ribosomal rna structures. Nucl. Acids Res. (2008)
4. Apostolico, A., Parida, L.: Incremental paradigms of motif discovery. J. Comput. Biol. **11**(1), 15–25 (2004)
5. Apostolico, A., Bock, M.E., Lonardi, S.: Monotony of surprise and large-scale quest for unusual words. J. Comput. Biol. **10**(2/3), 283–311 (2003)
6. Apostolico, A., Parida, L., Rombo, S.E.: Motif patterns in 2D. Theor. Comput. Sci. **390**(1), 40–55 (2008)
7. Bader, G., Hogue, H.: An automated method for finding molecular complexes in large protein-protein interaction networks. BMC Bioinform. **4**(2) (2003)
8. Bandyopadhyay, S., Sharan, R., Ideker, T.: Systematic identification of functional orthologs based on protein network comparison. Genome Res. **16**(3), 428–435 (2006)
9. Berg, J., Lassig, M.: Local graph alignment and motif search in biological networks. Proc. Natl. Acad. Sci. USA **101**(41), 14689–14694 (2004)
10. Bruckner, S., Hüffner, F., Karp, R.M., Shamir, R., Sharan, R.: Torque: topology-free querying of protein interaction networks. Nucl. Acids Res. **37**(Web-Server-Issue), 106–108 (2009)
11. Cheng, C.Y., Huang, C.Y., Sun, C.T.: Mining bridge and brick motifs from complex biological networks for functionally and statistically significant discovery. IEEE Trans. Syst. Man Cybern. Part B **38**(1), 17–24 (2008)
12. Ciriello, G., Guerra, C.: A review on models and algorithms for motif discovery in protein-protein interaction network. Brief. Funct. Genomics Proteomics (2008)
13. Ciriello, G., Mina, M., Guzzi, P.H., Cannataro, M., Guerra, C.: AlignNemo: A local network alignment method to integrate homology and topology. PLOS One **7**(6), e38,107 (2012)
14. Denielou, Y.P., Boyer, F., Viari, A., Sagot, M.F.: Multiple alignment of biological networks: a flexible approach. In: CPM'09 (2009)
15. Derenyi, I., Palla, G., Vicsek, T.: Clique percolation in random networks. Phys. Rev. Lett. **94**(16), 160–202 (2005)
16. Dost, B., et al.: Qnet: a tool for querying protein interaction networks. In: RECOMB'07, pp. 1–15 (2007)
17. Enright, A., Dongen, S., Ouzounis, C.: An efficient algorithm for large-scale detection of protein families. Nucl. Acids Res. **30**(7), 1575–84 (2002)
18. Erdos, P., Renyi, A.: On the evolution of random graphs. Publ. Math. Inst. Hung. Acad. Sci. **5**, 17–61 (1960)
19. Ferraro, N., Palopoli, L., Panni, S., Rombo, S.E.: Master-slave biological network alignment. In: 6th International symposium on Bioinformatics Research and Applications (ISBRA 2010), pp. 215–229 (2010)
20. Ferraro, N., et al.: Asymmetric comparison and querying of biological networks. IEEE/ACM Trans. Comput. Biol. Bioinform. **8**, 876–889 (2011)
21. Ferro, A., et al.: Netmatch: a cytoscape plugin for searching biological networks. Bioinformatics (2007)
22. Fionda, V., Palopoli, L., Panni, S., Rombo, S.E.: Protein-protein interaction network querying by a "focus and zoom" approach. In: BIRD'08, pp. 331–346 (2008)
23. Fionda, V., Panni, S., Palopoli, L., Rombo, S.E.: A technique to search functional similarities in PPI networks. Int. J. Data Mining Bioinform. (To appear)
24. Flannick, J., Novak, A., Graemlin, S., et al.: General and robust alignment of multiple large interaction networks. Genome Res. **16**(9), 1169–1181 (2006)
25. Fortunato, S.: Community detection in graphs. Phys. Rep. **486**, 75–174 (2010)
26. Furfaro, A., Groccia, M.C., Rombo, S.E.: Image classification based on 2D feature motifs. In: Flexible Query Answering Systems (FQAS 2013) (2013)

27. Garey, M., Johnson, D.: Computers and Intractability: A Guide to the Theory of NP-Completeness. Freeman, New York (1979)
28. Grossi, R., Pisanti, N., Crochemore, M., Sagot, M.F.: Bases of motifs for generating repeated patterns with wild cards. IEEE/ACM Trans. Comput. Biol. Bioinform. 2(3), 159–177 (2000)
29. Jancura, P., et al.: A methodology for detecting the orthology signal in a PPI network at a functional complex level. BMC Bioinform. (2011)
30. Kalaev, M., Bafna, V., Sharan, R.: Fast and accurate alignment of multiple protein networks. In: RECOMB'08 (2008)
31. Kelley, B., Yuan, B., Lewitter, F., Sharan, R., Stockwell, B.R., Ideker, T.: Pathblast: a tool for alignment of protein interaction networks. Nucl. Acid Res. 32, W83–W88 (2004)
32. Kim, W., Li, M., Wang, J., Pan, Y.: Biological network motif detection and evaluation. BMC Syst. Biol. 5(Suppl 3), S5 (2011)
33. King, A.D., Pržulj, N., Jurisica, I.: Protein complex prediction via cost-based clustering. Bioinformatics 20(17), 3013–3020 (2004)
34. Klau, G.W.: A new graph-based method for pairwise global network alignment. BMC Bioinform. 10(Suppl. 1), S59 (2009)
35. Koyuturk, M., Kim, Y., Topkara, U., Subramaniam, S., Szpankowski, W., Grama, A.: Pairwise alignment of protein interaction networks. J. Comput. Biol. 13(2), 182–199 (2006)
36. Kuchaiev, O., Przulj, N.: Integrative network alignment reveals large regions of global network similarity in yeast and human. Bioinformatics 27(10), 1390–1396 (2011)
37. Lacroix, V., Fernandes, C.G., Sagot, M.F.: Motif search in graphs: application to metabolic networks. IEEE/ACM Trans. Comput. Biol. Bioinform. 3(4), 360–368 (2006)
38. Leskovec, J., Lang, K., Mahoney, M.: Empirical comparison of algorithms for network community detection. In: Proceedings of the International World Wide Web Conference (WWW), pp. 631–640 (2010)
39. Liao, C.S., et al.: Isorankn: spectral methods for global alignment of multiple protein networks. Bioinformatics 25, i253–i258 (2009)
40. Madeira, S.C., Oliveira, A.L.: Biclustering algorithms for biological data analysis: a survey. IEEE Trans. Comput. Biol. Bioinform. 1(1), 24–45 (2004)
41. Mangan, S., Alon, U.: Structure and function of the feed-forward loop network motif. Proc. Natl. Acad. Sci. USA 100(21), 11980–11985 (2003)
42. Mangan, S., Itzkovitz, S., Zaslaver, A., Alon, U.: The incoherent feed-forward loop accelerates the response-time of the gal system of escherichia coli. J. Mol. Biol. 356(5), 1073–1081 (2005)
43. Milo, R., et al.: Network motifs: simple building blocks of complex networks. Science 298(5594), 824–827 (2002)
44. Mongiov, M., Sharan, R.: Global alignment of protein-protein interaction networks. In: Mamitsuka, H., DeLisi, C. Kanehisa, M. (eds.) Data Mining for Systems Biology, Methods in Molecular Biology, vol. 939, pp. 21–34. Humana Press (2013)
45. Neyshabur, B., Khadem1, A., Hashemifar, S., Arab, S.S.: NETAL: a new graph-based method for global alignment of protein?protein interaction networks. Bioinformatics 29(13), 11,654–1662 (2013)
46. Palla, G., et al.: Uncovering the overlapping community structure of complex networks in nature and society. Nature 435, 814–818 (2005)
47. Parida, L.: Pattern Discovery in Bioinformatics. Theory and Algorithms. Chapman and HAll/CRC (2008)
48. Parida, L.: Discovering topological motifs using a compact notation. J. Comput. Biol. 14(3), 46–69 (2007)
49. Park, Y., Song, M.: A genetic algorithm for clustering problems. In: Proceedings of 3rd Annual Conference on Genetic Algorithms, pp. 2–9 (1989)
50. Pinter, R., et al.: Alignment of metabolic pathways. Bioinformatics 21(16), 3401–3408 (2005)
51. Pizzuti, C., Rombo, S.E.: Experimental evaluation of topological-based fitness functions to detect complexes in PPI networks. In: Genetic and Evolutionary Computation Conference (GECCO), pp. 193–200 (2012)

52. Pizzuti, C., Rombo, S.E.: Multi-functional protein clustering in PPI networks. In: Proceedings of the 2nd International Conference on Bioinformatics Research and Development (BIRD), pp. 318–330 (2008)
53. Pizzuti, C., Rombo, S.E.: Pincoc: a co-clustering based approach to analyze protein-protein interaction networks. In: Proceedings of the 8th International Conference on Intelligent Data Engineering and Automated Learning, pp. 821–830 (2007)
54. Pizzuti, C., Rombo, S.E.: Restricted neighborhood search clustering revisited: an evolutionary computation perspective. In: Proceedings of the 8th IAPR International Conference on Pattern Recognition in Bioinformatics (PRIB), pp. 59–68 (2013)
55. Pizzuti, C., Rombo, S.E.: A coclustering approach for mining large protein-protein interaction networks. IEEE/ACM Trans. Comput. Biol. Bioinform. **9**(3), 717–730 (2012)
56. Rombo, S.E.: Optimal extraction of motif patterns in 2D. Inf. Process. Lett. **109**(17), 1015–1020 (2009)
57. Rombo, S.E.: Extracting string motif bases for quorum higher than two. Theor. Comput. Sci. **460**, 94–103 (2012)
58. Ruan, J., Zhang, W.: Identifying network communities with a high resolution. Phys. Rev. E **77**(1) (2008)
59. Sharan, R., et al.: From the cover: conserved patterns of protein interaction in multiple species. Proc. Natl. Acad. Sci. USA **102**(6), 1974–1979 (2005)
60. Sharan, R., Ideker, T.: Modeling cellular machinery through biological network comparison. Nat. Biotechnol. **24**(4), 427–433 (2006)
61. Shen-Orr, S.S., Milo, R., Mangan, S., Alon, U.: Network motifs in the trascriptional regulation network of escherichia coli. Nature **31**, 64–68 (2002)
62. Shih, Y.K., Parthasarathy, S.: Scalable global alignment for multiple biological networks. BMC Bioinform. **13**(Suppl 3), S11 (2012)
63. Shlomi, T., et al.: Qpath: a method for querying pathways in a protein-protein interaction network. BMC Bioinform. **7** (2006)
64. Singh, R., Xu, J., Berger, B.: Global alignment of multiple protein interaction networks. In: PSB'08 (2008)
65. Singh, R., Xu, J., Berger, B.: Pairwise global alignment of protein interaction networks by matching neighborhood topology. In: Research in Computational Molecular Biology (RECOMB 2007), pp. 16–31 (2007)
66. Van Dongen, S.: Graph clustering via a discrete uncoupling process. SIAM J. Matrix Anal. Appl. **30**(1), 121–141 (2008)
67. Wu, X., Liu, Q., Jiang, R.: Align human interactome with phenome to identify causative genes and networks underlying disease families. Bioinformatics **25**(1), 98–104 (2009)
68. Yang, Q., Sze, S.H.: Saga: a subgraph matching tool for biological graphs. J. Comput. Biol. **14**(1), 56–67 (2007)
69. Yeger-Lotem, E., et al.: Network motifs in integrated cellular networks of transcriptionregulation and proteinprotein interaction. Proc. Natl. Acad. Sci. USA **101**(16), 5934–5939 (2004)

Part II
Pattern Mining

This second part is devoted to presenting the pattern mining problem. We start by describing the early works on frequent pattern mining which are related to market basket analysis. We, then, introduce the problem of finding patterns able to discriminate between contrast populations (male/female, healthy/unhealthy, etc.) in the classical scenario where individuals are objects of a dataset. Subsequently, we introduce the problem of finding patterns on networks and on biological data which is the main topic of the present work. Finally, we present a novel challenging problem with a related solving proposal concerning discriminative pattern mining on biological networks.

Chapter 3
Exceptional Pattern Discovery

Abstract This chapter is devoted to a discussion on exceptional pattern discovery, namely on scenarios, contexts, and techniques concerning the mining of patterns which are so rare or so frequent to be considered as exceptional and, then, of interest for an expert to shed lights on the domain. Frequent patterns have found broad applications in areas like association rule mining, indexing, and clustering [1, 20, 23]. The application of frequent patterns in classification also achieved some success in the classification of relational data [6, 13, 14, 19, 25], text [15], and graphs [7]. The part is organized as follows. First, the frequent pattern mining on classical datasets is presented. This is not directly related with the content of the present work, which is mainly oriented in finding discriminating patterns, but they represent the starting point. Subsequently, Sect. 3.2 describes scenarios where patterns are exploited to discriminate between populations. Sections 3.3 and 3.4 illustrate how to mine patterns on networks and on biological data, respectively.

Keywords Frequent pattern · Emerging pattern · Interesting pattern · Network pattern · Pattern discovery · Subgraph retrieval

3.1 Frequent Pattern Mining

The earlier works about frequent pattern mining are in the field of market basket analysis with the seminal work [1] and the many variants that have been developed. The aim is to find frequent itemsets in an efficient way. At the basis of the technique there is the apriori property: *if an itemset is not frequent, each of its supersets is not frequent*. Exploiting this property, the huge search space can be pruned. Indeed, an itemset (and each superset) should be checked for being frequent if and only if each of its subsets is frequent, and can be pruned otherwise. The algorithm employs a breadth-first research approach. Often, a basic approach can lead to a set of redundant results, so several notions of maximal patterns have been defined in the literature to compress the output [16] or closed patterns [41].

© The Author(s) 2017   23
F. Fassetti et al., *Discriminative Pattern Discovery on Biological Networks*,
SpringerBriefs in Computer Science, DOI 10.1007/978-3-319-63477-7_3

3.2   Emerging Patterns and Contrast Sets

The frequent pattern mining, previously sketched, is aimed at finding significative patterns that could, in some way, represent interesting knowledge hidden in the analyzed population. However, in some contexts the problem is rather different. There are two populations to be analyzed and the knowledge we want to mine is the identification of peculiarities or characteristics holding in one population and not holding in the other one, namely characteristics that can discriminate individuals belonging to the populations at hand.

The earliest approaches in this scenario concern the emerging patterns [8, 28, 43]. An *emerging pattern* is a pattern whose frequency in one subpopulation drastically differs from its frequency in the other subpopulation. The growth rate of a pattern is defined as the ratio between the frequencies of the pattern in the two populations. If such a rate is above a threshold, the pattern is labeled as emerging. There are many families of emerging patterns in literature. Among them, the *jumping emerging patterns* [23] are emerging patterns where the growth rate is infinite. Thus, by definition, a jumping emerging pattern is a pattern that occurs in one population and does not occur in the other one.

Another important family is that of *constrained emerging patterns* [9]. The definition of this kind of patterns makes use of two thresholds α and β. A pattern is considered as emerging if the frequency in one population is lower than α and the frequency in the other population is greater than β.

Emerging patterns are also been employed to mine knowledge from gene expression data. In particular, based on emerging pattern mining algorithms and an entropy-oriented discretization method, [24] proposes a technique for retrieving a group of genes correlated to disease states in a statistically significant way. They found that some group of genes occurs in only one type of cells with an expression value ranging in a certain interval.

Besides emerging patterns, studies have been developed concerning contrast sets. Mining contrast sets refer to the task of understanding the differences between several contrasting groups and contrast sets are defined as conjunctions of attributes and values that differ meaningfully in their distribution across groups [11]. Smart algorithms aimed at pruning the search space have been introduced.

Differently from emerging patterns which are originally defined on the basis of the gap in support between sets, contrast sets are based on statistical consideration and the confidence notion. A contrast set can be seen as the antecedent of a rule $X \leftarrow Y$ that has a high confidence in one set and low confidence in the other set. In this scenario, the confidence is the probability $P(Y|X)$ of observing Y when X is observed and a potential contrast set X is discarded if it fails a statistical test for independence with respect to Y.

Contrast sets have been subsequently employed for classification purposes. Indeed, once the contrast sets have been identified they can be effectively employed to classify new individuals as belonging to one of the two sets [32].

Other related tasks are Subgroup Discovery [20], Change Mining [25], and others [28].

Another important family of techniques aimed at distinguish between groups is those of *outlier explanation*. This problem is somehow orthogonal to that of individuating outliers in a database. In this latter case, indeed, given characteristics to be analyzed, the aim is to find those individuals that are significantly dissimilar from the other individuals of the data population they belong to as far as that characteristics are concerned.

Conversely, in outlier explanation, the abnormal individual is given in advance, and it is of interest singling out the characteristics that best distinguish this individual from the rest of the data population.

It is assumed that one is given a significantly large data population characterized by a certain number of attributes, and information are provided that one (or a few) of the individuals in that data population is abnormal, but no reason whatsoever is given as to why this particular individual is to be considered abnormal. The problem we deal with here is precise to single out such reasons.

The problem can be tackled under several directions, since, due to its peculiarities, it needed the designing of specific techniques on the basis of one [3–5] or more outliers [6] in input and on the basis of the presence of numerical [3, 6] or categorical [4–6] attributes.

The problem is substantially different from emerging/contrast set mining, since here one of the set consists in one (or a few) individual and then ad-hoc technique is needed because no significant statistics can be extracted from the abnormal individuals.

3.3 Pattern Discovery on Networks

In this section the techniques aimed at mining frequent patterns on graphs are surveyed. In this context, a set of graphs is provided and patterns are defined as subgraphs, namely as a set of nodes and a set of complex binary relations between them, which can be frequently observed in the graphs of the set.

Previously, we have sketched methods for retrieving patterns on datasets where each individual is associated with a set of values on some fixed attributes.

Most of the proposed techniques are not interested in contrasting sets but they focus on finding patterns that can embed relevant knowledge about the population at hand.

Three are the more challenging problems: (i) find a pattern typically involves the subgraph isomorphism problem which is NP-hard; (ii) the number of mined patterns tends to be huge since pattern inclusion, maximal patterns, and subsumed patterns are nontrivial notions; (iii) the patterns found should be significant for the analyst.

Many efforts have been made in literature for solving these problems. In particular, some techniques hardly work on the first problem, trying to define searching strategy and pruning rules aimed at speeding up the mining process; other techniques are mainly focused on the search for significant patterns and then on defining notions of maximality or monotonicity to isolate interesting patterns.

The approach in [40] proposes the algorithm gSPAN (Graph-based Substructure PatterN mining). gSPAN adopts a depth-first search and a new labeling system for lexicographically ordering in graphs. Each graph is associated with a unique code and, exploiting these codes, a tree is built to model relations among graphs. The codes are also exploited to solve the isomorphism problem and the tree is analyzed to mine frequent patterns. Authors propose preprocessing and postprocessing technique to prune the search space and accelerate the mining process.

Authors of [17] propose a technique, called SPIN (SPanning tree-based maximal graph mINing), for efficiently mining maximal frequent subgraphs. In particular, they work on graphs having labeled nodes and edges, so the subgraph isomorphism problem and the maximality notion are simpler than the general case. The technique is based on mining all frequent tree patterns and then mining the maximal frequent subgraphs from trees. This allows to obtain good performance without a sensible loss of accuracy in the results.

In the context of mining patterns from graphs, the authors in [33] tackle the problem of retrieving just significant patterns. They provide a twofold contribution: an interesting technique to mine all frequent subgraphs and a method to identify significant patterns. The technique, in its part of mining frequent subgraphs, is an alternative to gSPAN with similar performance. The proposed method identifies a set of features (nodes, edges, or small subgraphs) and translates a graph in a feature vector through a random walk-based technique. Next, the p value of the feature vectors is computed and the subgraphs with the smallest p value are returned. Authors introduce also a notion of monotonicity about feature vector and super feature vector to avoid the retrieving of large patterns with the same characteristics of included patterns.

Finally, authors of [18] propose three variants of gSPAN: ATW-gSPAN, AW-gSPAN, and UBW-gSPAN for finding significant pattern on graphs having a weight associated with their edges. The three techniques vary on the basis of the way to compute the weight of a subgraph. ATW-gSPAN is the average weight of the pattern; AW-gSPAN relies on the affinity weight which takes into account the number of nodes composing the pattern; UBW-gSPAN is based on the notion of utility. The utility is influenced by the weighted support and by the concept of share which is the ratio of the graph weight of the pattern and the total weight of the set of graphs.

3.4 Pattern Discovery on Biological Data

Modeling biological information by graphs able to represent both cellular components and their associations with respect to physical interactions, co-expression, participation in related biological processes, etc., proved to be useful in order to extract new knowledge from the available biological annotations. Several computational approaches have been proposed in the last few years in order to discover evolutive conservations among different organisms, extract functional protein modules, or predict the biological function of some not yet well-characterized cellular components. They usually involve the solution of problems such as the alignment of different graphs [7, 13, 29, 35], the clustering of biological networks [2, 14, 19, 30, 31], or the extraction of significant patterns across the same or different graphs [21, 22, 27, 38]. More recently, the attention is turning on better understanding how the possible variations of cellular components (e.g., genes) may influence the emergence and course of diseases [10, 37]. To this aim it is necessary to combine genotype and phenotype aspects in the analysis, and graphs may represent again a powerful model. For example in [15], the relationships between genes and diseases, in terms of genomic variations involved in human diseases, inherited in the germ line or acquired somatically, are suitably mapped leading to the construction of the *diseasome*, that is, a network of disorders and disease genes linked by known disorder-gene associations. The analysis performed by the authors of [15] surprisingly showed that the vast majority of disease genes are nonessential with respect to the network topology, and they are localized in the functional periphery of the network. The authors provided explanation of their result by observing that only mutations compatible with survival are likely to be maintained in a population; therefore, disease-related mutations in the functionally and topologically peripheral regions of the cell give a higher chance of viability. On the other hand, they discussed how disease genes whose mutations are somatic should not be subject to such a selective pressure, indeed they found both functional and topological centrality for somatic cancer genes, confirming the hypothesis that many cancer genes play critical roles in cellular development and growth. This is in accordance with the *tumor bottleneck hypothesis* developed by [12]: If different genetic events contribute to a relatively uniform disease phenotype, their effect must eventually converge to a single gene or a small number of genes within the context of the tumor-driving cellular network. In this way, even if individual tumors of a particular type exhibit variation at the genomic level, they should all exhibit aberrant activity in a common set of regulators, which are typically situated within highly connected regulatory modules that lie upstream of the programs implementing and maintaining tumor cell homeostasis.

The findings mentioned above underline two important points: (1) The choice of a suitable model able to compactly represent the associations/interactions among cellular components is a key aspect to infer information at the phenotype level. (2) The changes induced by the emergence of a disorder or a disease at the genotype level are usually not clearly visible by exploring some cumulative model associated to a population. On the contrary, they involve slight variations which may also occur only in a part of the population and sometimes with some differences.

Building on these observations, we focused our attention on the identification of discriminative patterns, i.e., significant subgraphs characterizing the differences between two classes of graph sample sets (e.g., healthy and unhealthy individuals), where each graph is associated to an individual. Discriminative pattern mining has already been used in the bioinformatics context for several applications (see, e.g., [26]), but yet relatively few approaches relying on subgraph extraction have been proposed [34, 36, 39, 42]. In the following we provide some details on these latter approaches, all based on different notions of graph patterns aiming at identifying discriminative features characterizing the input datasets.

In [34] an approach is proposed for mining discriminative subgraphs from graphs with multiple labels. In particular, subgraphs that satisfy some criteria (e.g., the maximum substructure size, the minimum support, etc.) are enumerated and a multitarget regression model is built allowing to simultaneously collecting those subgraphs which can be considered *discriminative* to perform classification tasks. The approach has been applied to different datasets, among which a dataset of chemical compounds (drugs) with the goal of extracting subgroups triggering similar side effects.

In [36] the notion of *minimal contrast subgraph pattern* is introduced in order to single out structural differences between two collections of graphs. In particular, a contrast subgraph is a subgraph appearing in one class of graphs, but never in another class of graphs. It is also minimal if none of its subgraphs are contrasts. The authors of [36] propose an efficient approach based on the solution of the maximal common edge sets problem, and they applied it for the comparison of chemical compounds and to build graph classification models. Improved algorithms for the extraction of this kind of patterns were proposed in [42].

Synergy graph patterns have been defined in [38] by referring to subgraphs such that the relationships among the nodes are highly inseparable. The authors of [38] use confidence values to calculate discriminating power scores of graph patterns and consider only those graph patterns whose discriminating powers are much higher than all their subgraphs. They apply a classification algorithm based on synergy graph patterns to real-life datasets such as an AIDS antiviral screen chemical compounds dataset and anticancer screen datasets.

The authors of [39] consider the bioassay records for anticancer screen tests with different cancer cell lines and they build datasets belonging to a certain type of cancer screen with the outcome active or inactive. They propose an approach to distinguish these two classes based on the search of *dissimilar graph patterns*, according to a mining framework exploiting the correlation between structural similarity and significance similarity.

References

1. Agrawal, R., Srikant, R.: Fast algorithms for mining association rules in large databases. In: Proceedings of the 20th International Conference on Very Large Data Bases, VLDB'94, pp. 487–499. Morgan Kaufmann Publishers Inc., San Francisco, CA, USA (1994). http://dl.acm. org/citation.cfm?id=645920.672836
2. Ahn, Y., Bagrow, J., Lehmann, S.: Link communities reveal multiscale complexity in networks. Nature **466**, 761–764 (2010)
3. Angiulli, F., Fassetti, F., Manco, G., Palopoli, L.: Outlying property detection with numerical attributes. Data Min. Knowl. Discov. **31**(1), 134–163 (2017)
4. Angiulli, F., Fassetti, F., Palopoli, L.: Un metodo per la scoperta di proprietà inattese. In: SEBD, pp. 321–328 (2006)
5. Angiulli, F., Fassetti, F., Palopoli, L.: Detecting outlying properties of exceptional objects. ACM Trans. Database Syst. **34**(1) (2009)

6. Angiulli, F., Fassetti, F., Palopoli, L.: Discovering characterizations of the behavior of anomalous subpopulations. IEEE Trans. Knowl. Data Eng. **25**(6), 1280–1292 (2013)
7. Atias, N., Sharan, R.: Comparative analysis of protein networks: hard problems, practical solutions. Commun. ACM **55**(5), 88–97 (2012)
8. Bailey, J., Manoukian, T., Ramamohanarao, K.: Fast algorithms for mining emerging patterns. In: Proceedings of the 6th European Conference on Principles of Data Mining and Knowledge Discovery (PKDD), pp. 39–50. Springer-Verlag, London, UK (2002)
9. Bailey, J., Manoukian, T., Ramamohanarao, K.: Classification using constrained emerging patterns. In: Advances in Web-Age Information Management, pp. 226–237. Springer-Verlag (2003)
10. Barabasi, A.L., Gulbahce, N., Loscalzo, J.: Network medicine: a network-based approach to human disease. Nat. Rev. Genet. **12**(1), 56–68 (2011)
11. Bay, S.D., Pazzani, M.J.: Detecting group differences: mining contrast sets. Data Min. Knowl. Discov. **5**(3), 213–246 (2001)
12. Chen, J.C., Alvarez, M.J., Talos, F., et al.: Identification of causal genetic drivers of human disease through systems-level analysis of regulatory networks. Cell **159**(2), 402–414 (2014)
13. Ferraro, N., et al.: Asymmetric comparison and querying of biological networks. IEEE/ACM Trans. Comput. Biol. Bioinform. **8**, 876–889 (2011)
14. Georgii, E., et al.: Enumeration of condition-dependent dense modules in protein interaction networks. Bioinformatics **25**(7), 933–940 (2009)
15. Goh, K.I., Cusick, M.E., Valle, D., Childs, B., Vidal, M., Barabasi, A.L.: The human disease network. Proc. Natl. Acad. Sci. USA **104**(21), 8685–8690 (2007)
16. Gouda, K., Zaki, M.J.: Genmax: An efficient algorithm for mining maximal frequent itemsets. Data Min. Knowl. Discov. **11**(3), 223–242 (2005). doi:10.1007/s10618-005-0002-x
17. Huan, J., Wang, W., Prins, J., Yang, J.: Spin: Mining maximal frequent subgraphs from graph databases. In: Proceedings of the ACM SIGKDD International Conference on Knowledge Discovery and Data Mining (KDD), pp. 581–586 (2004)
18. Jiang, C., Coenen, F., Zito, M.: Frequent sub-graph mining on edge weighted graphs. In: Proceedings of the International Conference on Data Warehousing and Knowledge Discovery (DAWAK), pp. 77–88 (2010)
19. Jiang, P., Singh, M.: SPICi: a fast clustering algorithm for large biological networks. Bioinformatics **26**(8), 1105–1111 (2010)
20. Klösgen, W.: Explora: A multipattern and multistrategy discovery assistant. In: Advances in Knowledge Discovery and Data Mining (KDD), pp. 249–271 (1996)
21. Koyuturk, M., Grama, A., Szpankowski, W.: An efficient algorithm for detecting frequent subgraphs in biological networks. Bioinformatics **20**(1), 200–207 (2004)
22. Koyutürk, M., Kim, Y., Subramaniam, S., Szpankowski, W., Grama, A.: Detecting conserved interaction patterns in biological networks. J. Comput. Biol. **13**(7), 1299–1322 (2006)
23. Li, J., Dong, G., Ramamohanarao, K.: Making use of the most expressive jumping emerging patterns for classification. Knowl. Inf. Syst. **3**(2), 1–29 (2001)
24. Li, J., Wong, L.: Emerging patterns and gene expression data. Genome Inf. **12**, 3–13 (2001)
25. Liu, B., Hsu, W., Ma, Y.: Discovering the set of fundamental rule changes. In: Proceedings of the ACM SIGKDD International Conference on Knowledge Discovery and Data Mining (KDD), pp. 335–340 (2001)
26. Liu, X., Wu, J., Gu, F., Wang, J., He, Z.: Discriminative pattern mining and its applications in bioinformatics. Brief. Bioinform. **16**(5), 884–900 (2015)
27. Milo, R., et al.: Network motifs: Simple building blocks of complex networks. Science **298**(5594), 824–827 (2002)
28. Novak, P.K., Lavrac, N., Webb, G.I.: Supervised descriptive rule discovery: a unifying survey of contrast set, emerging pattern and subgroup mining. J. Mach. Learn. Res. **10**, 377–403 (2009)
29. Panni, S., Rombo, S.E.: Searching for repetitions in biological networks: methods, resources and tools. Brief. Bioinform. **16**(1), 118–136 (2015)
30. Pizzuti, C., Rombo, S.E.: Algorithms and tools for protein-protein interaction networks clustering, with a special focus on population-based stochastic methods. Bioinformatics **30**(10), 1343–1352 (2014)

31. Pizzuti, C., Rombo, S.E., Marchiori, E.: Complex detection in protein-protein interaction networks: a compact overview for researchers and practitioners. In: 10th European Conference of Evolutionary Computation, Machine Learning and Data Mining in Bioinformatics (EvoBio), pp. 211–223 (2012)
32. Ramamohanarao, K., Bailey, J., Fan, H.: Efficient mining of contrast patterns and their applications to classification. In: Proceedings of the International Conference on Intelligent Sensing and Information Processing (ICISIP), pp. 39–47 (2005)
33. Ranu, S., Singh, A.K.: Graphsig: a scalable approach to mining significant subgraphs in large graph databases. In: Proceedings of the IEEE International Conference on Data Engineering, pp. 844–855 (2009)
34. Shao, Z., Hirayama, Y., Yamanishi, Y., Saigo, H.: Mining discriminative patterns from graph data with multiple labels and its application to quantitative structure-activity relationship (QSAR) models. J. Chem. Inf. Model. **55**(12), 2519–2527 (2015)
35. Singh, R., Xu, J., Berger, B.: Isorank: global alignment of multiple protein interaction networks with applications to functional orthology detection. Proc. Natl. Acad. Sci. **105**(35), 12763–12768 (2008)
36. Ting, R.M.H., Bailey, J.: Mining minimal contrast subgraph patterns. In: SIAM International Conference on Data Mining (SDM) (2006)
37. Vidal, M., Cusick, M.E., Barabasi, A.L.: Interactome networks and human disease. Cell **144**(6), 986–998 (2011)
38. Wang, Z., Zhao, Y., Wang, G., Li, Y., Wang, X.: On extending extreme learning machine to non-redundant synergy pattern based graph classification. Neurocomputing **149, Part A**(0), 330–339 (2015)
39. Yan, X., Cheng, H., Han, J., Yu, P.S.: Mining significant graph patterns by leap search. In: ACM SIGMOD International Conference on Management of data, pp. 433–444. ACM (2008)
40. Yan, X., Han, J.: gspan: Graph-based substructure pattern mining. In: Proceedings of the IEEE International Conference on Data Mining (ICDM), pp. 721–724 (2002)
41. Zaki, M.J., Hsiao, C.J.: Charm: An efficient algorithm for closed itemset mining. In: Proceedings of the SIAM International Conference on Data Mining (SDM), pp. 457–473 (2002)
42. Zeng, Z., Wang, J., Zhou, L.: Efficient mining of minimal distinguishing subgraph patterns from graph databases. In: Advances in Knowledge Discovery and Data Mining, pp. 1062–1068 (2008)
43. Zhang, X., Dong, G., Kotagiri, R.: Exploring constraints to efficiently mine emerging patterns from large high-dimensional datasets. In: Proceedings of the Sixth ACM SIGKDD International Conference on Knowledge Discovery and Data Mining, KDD'00, pp. 310–314. ACM, New York, NY, USA (2000). doi:10.1145/347090.347158

Chapter 4
Discriminating Graph Pattern Mining from Gene Expression Data

Abstract Here we consider the problem of mining gene expression data in order to single out interesting features characterizing healthy/ unhealthy samples of an input dataset. The presented approach is based on a network model of the input gene expression data, where there is a labeled graph for each sample. This is the first attempt to build a different graph for each sample and, then, to have a database of graphs for representing a sample set. The main goal is that of singling out interesting differences between healthy and unhealthy samples, through the extraction of discriminative patterns among graphs belonging to the two different sample sets. Differently from the other approaches presented in the literature, this technique is able to take into account important local similarities, and also collaborative effects involving interactions between multiple genes. In particular, edge-labeled graphs are employed and the discriminative power of a pattern is measured on the basis of edge weights, which are representative of how much relevant is the co-expression between two genes.

Keywords Gene expression data · Network model · Discriminating pattern · Subgraph discovery · Pattern relevance · Correlation measure

4.1 Motivations

Mechanisms regulating the organization and functioning of cells are still not completely understood, although it is commonly recognized that they are based on the interplay of several different factors. For many decades, single molecules playing important roles in the cell, such as proteins, genes, and RNA, have been deeply studied as independent objects. At the beginning of this century, after that the genome sequencing of many organisms, among which human have been completed, the attention has turned on how cellular components interact each other in order to accomplish together specific biological functions. Now that the next-generation sequencing techniques allow to obtain accurate and abundant data at the cellular level, great interest is emerging from the analysis of genotype–phenotype relationships in order to understand their connection with the course of diseases. In this scenario, suitable models

© The Author(s) 2017   31
F. Fassetti et al., *Discriminative Pattern Discovery on Biological Networks*,
SpringerBriefs in Computer Science, DOI 10.1007/978-3-319-63477-7_4

may be usefully adopted to answer many unsolved questions about biological systems and their collective functioning; this is the first step to throw light on complex relationships between genotype and phenotype in order to analyze the molecular basis of diseases.

The complex interactions occurring within a cell may be modeled by biological networks, including gene regulatory networks, gene co-expression networks, protein–protein interaction networks, and metabolic networks [3, 5, 9, 13]. For instance, protein–protein interaction networks represent pairwise interactions between proteins, whereas metabolic networks model the chemical pathways occurring in metabolic reactions. Building all these kinds of networks is possible, thanks to the information stored in public interaction databases and mainly obtained by high-throughput technologies. Here we focus on networks based on gene expression data, and we recall some basic notions below on this type of data (the interested reader can find exhaustive information on the use of genome-wide gene expression data at [12]).

The transcriptome of a cell comprises mRNA, tRNA, rRNA, and short regulatory RNAs. In order to generate large-scale gene expression data, biologists use microarray experiments, that is, they measure genome-wide gene expression levels of mRNA in a cell or a tissue sample under a particular condition. A microarray chip quantifies the hybridization of fluorescent labeled target nucleotide sequences to define complementary probe sequences that are spotted on a glass or silicon slide. More details about microarrays may be found in [1, 4, 10].

In the last few years, also more sophisticated techniques have been developed such as next-generation sequencing (RNA-seq) [7, 14]. RNA-seq has a wide variety of applications such as the measurement of gene expression levels from transcribed mRNA sequences.

All these technologies have revolutionized the biological research but it is challenging to interpret the direct results from experiments to investigate complex biological mechanisms. Therefore, a number of techniques have been proposed to switch from a tabular to a network representation (see [11] to have some references about them).

Traditional techniques start from microarray measurements to find out the expression level of each gene in each of the samples analyzed; this data is used to define the so-called "profile expression" of each gene over the sample set. Statistical, machine learning, or soft-computing techniques have been introduced for the co-expression networks construction, but all of them need to look at the sample set globally.

Actually, it has been observed that the expression profiles often share local rather than global similarities [11], so if one tries to model cellular mechanisms of an organism through a graph, some potentially powerful details of each interaction may be left aside.

Here we consider the problem of identifying interesting differences between two input sample sets, associated to healthy and unhealthy individuals, respectively. To this aim, we propose an approach based on two main characteristics: (i) a representation of gene co-expression data able to take into account local similarities, and (ii) the definition of a suitable notion of pattern useful to capture the differences between the two input sample sets. In particular, our model emphasizes the importance of locality by turning the microarray dataset into a graph dataset, where there is a labeled graph for each sample. Note that, in this context, the number of samples is much smaller than the number of genes. The main aim of our approach is that of singling out interesting differences between healthy and unhealthy samples, through the extraction of "discriminating patterns" among graphs belonging to the two different sample sets.

It is worth to point out that common discriminating graph pattern mining approaches have been shown to achieve great success by mining the graph patterns that occur with disproportionate frequency in some classes versus others [16]. However, this kind of information may be not enough when mining biological graph patterns, especially if one wants to capture those interactions that can be related to a certain pathological phenotype. Indeed, diseases such as cancer are often related to collaborative effects involving interactions between multiple genes or proteins [2, 15]. Therefore, the discriminating power of a pattern should be higher than the one of all its sub-patterns. We contribute in this direction both by enumerating patterns with node labels, which are associated to how much *relevant* is the co-expression between two genes, and by introducing a measure of how much discriminating is a pattern, based on the edge weights.

4.2 Network Model

The basic input data considered here is *gene expression data*. Gene expression data contains information about the expression level of several genes on the set of analyzed individuals. They are represented as a multiset of tuples on a set of attributes, where each individual is associated with a tuple and each attribute is associated with a gene.[1] The value $t(a)$ that a tuple t assumes on an attribute a is the level of expression of the gene associated with a for the individual t.

Let (V, E) be a *labeled undirected graph* where V is a set of *nodes* (or *vertices*), each identified by a unique label, and E is a set of *edges*, i.e., unordered pairs (v, w) where $v, w \in V$. A sequence $v_1, \ldots, v_h$ of nodes in V such that $(v_i, v_{i+1}) \in E$ for any $i \in 1..(h-1)$ is a *path* from v_1 to v_h. We recall that a graph is *connected* if for each pair of nodes v and w, there is at least a path from v to w.

Given an input set of gene expression data, in our model each gene is associated to a node v of a labeled undirected graph. Each edge connecting a pair of genes has two different weights: (1) the *strength* of the relationship between these two genes, and (2) the *relevance*. The latter is used to make the analysis more robust to statistical fluctuations, since, roughly speaking, it represents the exceptionality of the strength associated with e, w.r.t. the expected value.

---

[1] Since there is a one-to-one correspondence between an individual and its representing tuple, for the sake of simplicity, we employ the same symbol t to denote both the individual and its corresponding tuple in the dataset.

In the following of the section, we first formally introduce the notions of strength and relevance and, then, we formalize the adopted network model.

4.2.1 Strength of the Relationships

Given a population **DS**, an individual t of **DS** and two genes a_i and a_j, we aim at relating the *strength* of the relationship between a_i and a_j for t to the correlation ρ^t_{ij} between $t(a_i)$ and $t(a_j)$. A first problem is to estimate the correlation between $t(a_i)$ and $t(a_j)$, for each tuple t and for any pair of attributes a_i and a_j. Note that, despite classical approaches, the correlation considered here is based on only two observations.

Let X^t_i (X^t_j, resp.) be the random variables associated with $t(a_i)$ ($t(a_j)$, resp.), and consider the bivariate normal distribution having mean vector μ_{ij} and covariance matrix Σ^t_{ij}, where

$$\mu_{ij} = \begin{pmatrix} \mu_i \\ \mu_j \end{pmatrix}, \ \Sigma^t_{ij} = \begin{pmatrix} \sigma^2_i & \rho^t_{ij}\sigma_i\sigma_j \\ \rho^t_{ij}\sigma_i\sigma_j & \sigma^2_j \end{pmatrix},$$

μ_i (μ_j, resp.) is the mean value of attribute a_i (a_j, resp.), σ_i (σ_j, resp.) is the standard deviation of attribute a_i (a_j, resp.), then independent from t, and ρ^t_{ij} is the correlation between X^t_i and X^t_j.

In order to emphasize the impact of the observed values, an interesting value of correlation $\rho^t_{i,j}$ between X^t_i and X^t_j can be estimated by inferring the value ρ^t_{ij} maximizing the probability of observing the two-dimensional point $[t(a_i), t(a_j)]$ and, thus, it can be suitably employed as strength of the relationship between them.

Definition 4.1 (*Strength*) Given a population **DS**, an individual t in **DS**, and two genes a_i and a_j, the *strength* of the relation between a_i and a_j for t is the value of correlation that maximizes the probability of observing $t(a_i)$ and $t(a_j)$.

Strength Computation

In order to ease the computation, we normalize each value according to the mean and the standard deviation of the associated attribute. Hence, we compute $\hat{t}_i = \frac{t(a_i)-\mu_i}{\sigma_i}$ and $\hat{t}_j = \frac{t(a_j)-\mu_j}{\sigma_j}$.

Let $\widehat{X}^t_i$ ($\widehat{X}^t_j$, resp.) be the random variable associated with $\hat{t}_i$ ($\hat{t}_j$, resp.) and consider the bivariate normal distribution with components $\widehat{X}^t_i$ and $\widehat{X}^t_j$, mean vector $\widehat{\mu}$, and covariance matrix $\widehat{\Sigma}$, where

$$\widehat{\mu}_{ij} = \begin{pmatrix} 0 \\ 0 \end{pmatrix}, \ \widehat{\Sigma}^t_{ij} = \begin{pmatrix} 1 & \rho_{t_i t_j} \\ \rho_{t_i t_j} & 1 \end{pmatrix}.$$

Thus, the bivariate normal distribution can be written as

$$f(x, y, \rho_{t_i t_j}) = \frac{1}{2\pi\sqrt{1 - \rho_{t_i t_j}^2}} e^{-\frac{1}{2\left(1-\rho_{t_i t_j}^2\right)}\left(x^2+y^2-2\rho_{t_i t_j}xy\right)}.$$

The aim is to find the value $\tilde{\rho}_{t_i t_j}$ of $\rho_{t_i t_j}$ such that the value of f in the point $(\hat{t}_i, \hat{t}_j, \rho_{t_i t_j})$ is maximum, in the following formula:

$$\tilde{\rho}_{t_i t_j} = \arg\max_{\rho_{t_i t_j}} f(\hat{t}_i, \hat{t}_j, \rho_{t_i t_j}),$$

which represents the strength between genes a_i and a_j for t. It can be proved[2] that the stationary points are obtainable as the solution of

$$\rho^3 - \rho^2 \hat{t}_i \hat{t}_j + \rho(\hat{t}_i^2 + \hat{t}_j^2 - 1) - \hat{t}_i \hat{t}_j = 0. \tag{4.1}$$

4.2.2 Relevance of the Relationships

In the previous section, we estimated how much two observations are correlated. In order to make this estimation more robust, we measure the probability that a possible high value of correlation is not due by chance. The underlying idea is to test the null hypothesis under which a high value of correlation could be implied by a certain value of expression of a gene for an individual. In other words, given a certain value of expression, it could be quite high the probability that the level of expression of another gene leads to a high value of correlation.

Definition 4.2 (*Relevance*) Given a population **DS**, an individual t of **DS**, and two genes a_i and a_j, the *relevance* of the relation between a_i and a_j for t is the minimum between the probability of observing a strength smaller than $\tilde{\rho}_{ij}^t$ given the level of expression of a_i equal to $t(a_i)$ and the probability of observing a strength smaller than $\tilde{\rho}_{ij}^t$ given the level of expression of a_j equal to $t(a_j)$.

Intuitively speaking, the higher the relevance the smaller is the probability that the observed value of correlation is due by chance.

Relevance Computation

Let t_i^* (t_j^*, resp.) be the observed expression level of gene a_i (a_j, resp.) in t and let ρ^* be the strength associated with t_i^* and t_j^*. Moreover, let P_i^- (P_j^-, resp.) be the probability of observing a strength smaller than ρ^* given $t(a_i) = t_i^*$ ($t(a_j) = t_j^*$, resp.). Then, the relevance between a_i and a_j for t is

$$\min(P_i^-, P_j^-) = 1 - \max(P_i^-, P_j^-),$$

where P_i^- and P_j^- can be rewritten as

---

[2]The reader is referred to Sect. 4.4.2 for the details.

$$P_i^- = 1 - Pr(\rho \geq \rho^* | t(a_i) = t_i^*) = 1 - P_i^+$$
$$P_j^- = 1 - Pr(\rho \geq \rho^* | t(a_j) = t_j^*) = 1 - P_j^+.$$

In order to evaluate the relevance, we can compute the probability P_i^+ (P_j^+, resp.) of observing a value of $t(a_i)$ ($t(a_j)$, resp.) such that the strength of a_i and a_j for t is greater than ρ^*, by keeping t_j^* (t_i^*, resp.) fixed.

Consider Eq. 4.1 again. By solving it with respect to $\hat{t}_i$ ($\hat{t}_j$, resp.) and by keeping ρ and $\hat{t}_j$ ($\hat{t}_i$, resp.) fixed, we can determine two points t_i', t_i'' such that the strength of a_i and a_j for t is greater that ρ^* for any $t_i' \leq (t(a_i) - \mu_i)/\sigma_i \leq t_i''$. Thus, the probabilities P_i^+ and P_j^+ can be computed as

$$P_i^+ = Pr(\widehat{X}_i \leq t_i'') - Pr(\widehat{X}_i \leq t_i') = \Phi(t_i'') - \Phi(t_i')$$
$$P_j^+ = Pr(\widehat{X}_j \leq t_j'') - Pr(\widehat{X}_j \leq t_j') = \Phi(t_j'') - \Phi(t_j'),$$

where $\Phi(\cdot)$ denotes the cumulative distribution function of the standard normal distribution.

4.2.3 Building Networks

In this section we tackle the problem of building a distinct network for each individual of a given population, so that the obtained database of graphs could be employed for the subsequent phase of mining. By enriching the classical graph model, we add two weights to each edge (strength and relevance) and obtain the following model of network.

Definition 4.3 (*SR-network*) Given a set of nodes V, a *SR-network* (standing for *Strength-Relevance-Network*) on V is a quadruple (V, E, η, π) where E is a set of edges, $\eta : E \rightarrow \Re$ is a function associating each edge $e \in E$ with a real number representing the *strength* of e, and $\pi : E \rightarrow [0, 1]$ is a function associating each edge in $e \in E$ with a real number between 0 and 1 representing the *relevance* of e.

For each individual t in **DS**, a *SR-network* $\mathcal{N}_t = (V, E_t, \eta_t, \pi_t)$ is associated with t and built as follows. Each gene a_i is associated to a node $v_i \in V$. For each pair of genes a_i and a_j, the edge $e(v_i, v_j)$ is inserted in E_t if and only if the relation between a_i and a_j is both strong and relevant, namely, if and only if the strength is larger than a threshold τ_s and the relevance is larger than a threshold τ_r.

Given a population **DS** consisting of m individuals, we can then obtain a database of m *SR-networks* $\{\mathcal{N}_1 = (V, E_1, \eta_1, \pi_1), \mathcal{N}_2 = (V, E_2, \eta_2, \pi_2), \ldots, \mathcal{N}_m = (V, E_m, \eta_m, \pi_m)\}$, where the *SR-network* $\mathcal{N}_i$ is associated with the i-th individual of **DS**.

4.3   Statement of the Problem

This section is devoted to formally introduce the main problem we are interested in solving.

Given a population **DS**, suppose that it is apriori partitioned in two groups $\mathbf{DS}_1$, $\mathbf{DS}_2$ on the basis of certain properties of the samples (i.e., healthy vs unhealthy).

The goal is to single out peculiarities of a subpopulation w.r.t. the other one. This can be exploited to shedding light on the characteristics that distinguish individuals of $\mathbf{DS}_1$ from individuals of $\mathbf{DS}_2$. Since this kind of knowledge is often related to collaborative effects involving interactions between multiple genes or proteins [2, 15], we aim at mining graph patterns that can be in charge of the separation between the subpopulations at hand. In particular, we search for patterns that are representative of one subpopulation, but not of the other one.

First, we introduce the notion of pattern which is the building block of the knowledge we want to mine.

Definition 4.4 (*Pattern*) Given a *SR-networks* database **N** defined on a set of nodes V, a *pattern* $\mathscr{P}$ in **N** is a connected graph (Vp, Ep) with $Vp \subseteq V$.

A pattern $\mathscr{P}' = (V_{\mathscr{P}'}, E_{\mathscr{P}'})$ is a sub-pattern of $\mathscr{P} = (V_{\mathscr{P}}, E_{\mathscr{P}})$ (or, equivalently $\mathscr{P}$ is a super-pattern of $\mathscr{P}'$) if $V_{\mathscr{P}'} \subseteq V_{\mathscr{P}}$ and $E_{\mathscr{P}'} \subseteq E_{\mathscr{P}}$.

Given a *SR-network* $\mathscr{N} = (V, E, \eta, \pi)$ and a pattern $\mathscr{P} = (V_P, E_P)$, there is a *match* of $\mathscr{P}$ in $\mathscr{N}$ if and only if $V_P \subseteq V$ and $E_P \subseteq E$.

Since, by construction, all networks and all patterns are defined on the same set of nodes and each node is different from each other being associated with a different gene, the following property clearly holds.

Property 4.1 *Given a database* **N** *and a pattern* $\mathscr{P}$ *in* **N**, *for any network* $\mathscr{N}$ *in* **N** *exists only one match of* $\mathscr{P}$ *in* $\mathscr{N}$.

Therefore, we say that a network $\mathscr{N}$ *matches* a pattern $\mathscr{P}$ (or, $\mathscr{P}$ *occurs in* $\mathscr{N}$), meaning that there is a match of $\mathscr{P}$ in $\mathscr{N}$, and such a match is unambiguously determined.

Next, we extend the notion of *strength* and *relevance* provided in Sects. 4.2.1 and 4.2.2 to a pattern $\mathscr{P}$ in a *SR-network* $\mathscr{N} = (V, E, \eta, \pi)$. In particular, the strength $\eta(\mathscr{P}, \mathscr{N})$ of the match of a pattern $\mathscr{P} = (V_P, E_P)$ in $\mathscr{N}$ is defined as

$$\eta(\mathscr{P}, \mathscr{N}) = \frac{1}{|E_P|} \cdot \sum_{e \in E_P} \eta(e),$$

while the relevance of $\mathscr{P}$ in $\mathscr{N}$ is defined as the product of the relevancies of the edges of $\mathscr{P}$ in $\mathscr{N}$

$$\pi(\mathscr{P}, \mathscr{N}) = \prod_{e \in E_P} \pi(e).$$

The ratio underlying these definitions is that, in order for a pattern to be relevant, all its edges should be relevant. Since the relevance is a sort of "probability of observing the edge," the product of the relevancies, similarly to the probability of intersection between events, takes into account this issue.

In order to evaluate how much a pattern is representative of a (sub)population, we need to compute how much a pattern is common to occur in that population. Note that simply counting the number of *SR-networks* in the database matching the pattern could be misleading, since all the information coming from strength and relevance is neglected. To obtain a robust measure, we evaluate the *commonness* of a pattern $\mathscr{P}$ in a (sub)population $\mathbf{N}$ and denote it by $s(\mathscr{P}, \mathbf{N})$, where only relevant matches are considered, and the strengths of the matches are summed:

$$s(\mathscr{P}, \mathbf{N}) = \sum_{\mathscr{N} \in \mathbf{N}: \pi(\mathscr{P}, \mathscr{N}) > \tau_r} \eta(\mathscr{P}, \mathscr{N}).$$

According to the statistical test theory, τ_r can be set to the standard values for testing hypotheses.

Property 4.2 *Given a database $\mathbf{N}$ and a pattern $\mathscr{P}$, the commonness of $\mathscr{P}$ in $\mathbf{N}$ is upper bounded by the support of $\mathscr{P}$ in $\mathscr{N}$, i.e., the number of SR-networks in $\mathbf{N}$ where $\mathscr{P}$ occurs.*

The property immediately follows from the observations that the strength of a pattern in a *SR-network* $\mathscr{N}$ ranges in $[0, 1]$ and that it is 0 if the pattern has no matches in $\mathscr{N}$, while the support is 0 if the pattern has no matches in $\mathscr{N}$ and it is 1 if the pattern occurs in $\mathscr{N}$.

4.3.1 Discriminating Pattern

In order to measure the discriminating power of a pattern we resort to the notion of information gain [8] and adapt it to our context. The aim is to measure the change in information entropy [6] led by the pattern.

Let $\mathbf{N}$ be a population partitioned in two subpopulations $\mathbf{N}_1$ and $\mathbf{N}_2$ and let $\mathscr{P}$ be a pattern. The *discriminating power* of $\mathscr{P}$, denoted as $pow(\mathscr{P})$, is the gain in entropy

$$pow(\mathscr{P}) = H(\mathbf{N}) - H(\mathbf{N}|\mathscr{P}),$$

namely, the difference between the entropy of the population $H(\mathbf{N})$ and the entropy $H(\mathbf{N}|\mathscr{P})$ of the population given the pattern.

Thus, in order to define the discriminating power, we need to adapt the notion of entropy to our context.

The *information entropy $H(\mathbf{N})$* is

$$H(\mathbf{N}) = -\frac{|\mathbf{N}_1|}{|\mathbf{N}|} \log \frac{|\mathbf{N}_1|}{|\mathbf{N}|} - \frac{|\mathbf{N}_2|}{|\mathbf{N}|} \log \frac{|\mathbf{N}_2|}{|\mathbf{N}|}.$$

As for the information entropy conditioned by the pattern $\mathcal{P}$, $H(\mathbf{N}|\mathcal{P})$, we note that the pattern $\mathcal{P}$ partitions the population $\mathbf{N}$ in two groups of individuals those where $\mathcal{P}$ is relevant, denoted as $\mathbf{N}^{\mathcal{P}}$, and those where the pattern $\mathcal{P}$ it not relevant, denoted as $\mathbf{N}^{\overline{\mathcal{P}}}$. The entropy of $\mathbf{N}$ conditioned by $\mathcal{P}$ can be, then, computed as

$$H(\mathbf{N}|\mathcal{P}) = H\left(\mathbf{N}^{\mathcal{P}}\right) \cdot q + H\left(\mathbf{N}^{\overline{\mathcal{P}}}\right) \cdot (1 - q)$$

with $q = \frac{s(\mathcal{P},\mathbf{N}_1)+s(\mathcal{P},\mathbf{N}_2)}{|\mathbf{N}|}$, and

$$H(\mathbf{N}^{\mathcal{P}}) = -q_1 \log q_1 - (1 - q_1) \log(1 - q_1)$$
$$H(\mathbf{N}^{\overline{\mathcal{P}}}) = -q_2 \log q_2 - (1 - q_2) \log(1 - q_2)$$

with

$$q_1 = \frac{s(\mathcal{P}, \mathbf{N}_1)}{s(\mathcal{P}, \mathbf{N}_1) + s(\mathcal{P}, \mathbf{N}_2)} \text{ and}$$
$$q_2 = \frac{|\mathbf{N}_1| - s(\mathcal{P}, \mathbf{N}_1)}{|\mathbf{N}_1| - s(\mathcal{P}, \mathbf{N}_1) + |\mathbf{N}_2| - s(\mathcal{P}, \mathbf{N}_2)}.$$

Next, we provide a formal definition of the particular type of patterns we are interested in.

The patterns that the algorithm must highlight are those ones that give more discriminating than their subgraphs (i.e., the discriminating power of the pattern should be higher than the one of all its sub-patterns). Moreover, if two patterns have the same discriminating power we take only the one with maximum commonness. Indeed, the discriminating power of a pattern that is low-supported can harm the accuracy of the analysis due to overfitting, and then also in this case we need to take into account the commonness.

Definition 4.5 (*Discriminating pattern*) A pattern $\mathcal{P}$ is a *discriminating* pattern if and only if for each pattern $\mathcal{P}'$ sub-pattern of $\mathcal{P}$ either $pow(\mathcal{P}) > pow(P')$ or $pow(\mathcal{P}) = pow(P')$ and $s(\mathcal{P}) > s(P')$.

This definition is relevant in biological fields as disease such as cancer is often related to collaborative effects involving interactions between multiple genes or proteins [2] [15], so those patterns whose discriminating power gives us more information than one of their sub-patterns seemed to be really interesting, as it means that the relations among genes in the pattern can give more information about the subpopulation than their subsets can do.

Algorithm 1: Discovering Discriminating Pattern

Input: Gene expression data **DS** (partitioned in DS_1 and in DS_2), thresholds τ_r and τ_s
Output: Discriminating patterns, *res*
$N_1 \leftarrow$ BUILDSRNETWORKS(DS_1, τ_r, τ_s)
$N_2 \leftarrow$ BUILDSRNETWORKS(DS_2, τ_r, τ_s)
foreach $N_{MAIN} \in \{N_1, N_2\}$ **do**
    $edges \leftarrow$ SORTEDGES(N_{MAIN})
    $res \leftarrow$ PATTERNMINE($\emptyset$, $edges$, $\emptyset$, τ_r, $\tau_s \cdot |N_{MAIN}|$)
Delete non-maximal pattern in res
return *res*

4.3.2 Problem Definition

Patterns generated according to this definition may be redundant. Indeed, suppose you have mined a pattern $\mathscr{P}$ which is discriminating since its discriminating power is higher than that of its sub-patterns. If there is a pattern $\mathscr{P}'$ super-pattern $\mathscr{P}$ such that $pow(\mathscr{P}') > pow(\mathscr{P})$, according to Definition 4.5, $\mathscr{P}'$ is a discriminating patterns as well. To avoid keeping both $\mathscr{P}$ and $\mathscr{P}'$ in the result set of discriminating patterns, we resort to a notion of maximality as formalized in the following definition.

Definition 4.6 (*Maximal discriminating pattern*) A discriminating pattern $\mathscr{P}$ is said to be maximal if and only if there is not a discriminating pattern $\mathscr{P}'$ such that $\mathscr{P}$ is a sub-pattern of $\mathscr{P}'$.

Thus, the problem we are interested in solving consists in singling out all the maximal discriminating patterns.

4.4 Technique

This section is devoted to present the algorithm we proposed to mine discriminating patterns, sketched in Algorithm 1.

4.4.1 Algorithm

We now present the algorithm proposed to mine discriminating patterns, sketched in Algorithm 1. The algorithm, basically, visits in depth and tries to prune the search space consisting in connected subgraphs of the input networks. In more details, for each subpopulation we build the associated database of *SR-networks* as described in Sect. 4.2. Next, we mine patterns in two phases by singling out patterns over represented in subpopulation DS_1 against DS_2 and then those over represented in subpopulation DS_2 against DS_1. In other words, we aim at detecting patterns which

---

Function PATTERNMINE($\mathscr{P}_{cur}$, $neighs$, $visited\,Edges$)

Input: Current Pattern [$\mathscr{P}_{cur}$]
          pattern neighborhood [$neighs$]
          already visited edges [vE]
          thresholds τ_r and T_s
Output: Discriminating pattern candidates [res]
foreach $edge$ **in** $neighs$ **do**
> $\mathscr{P}_{next} \leftarrow \mathscr{P}_{cur} \cup edge$
> $vE \leftarrow vE \cup edge$
> $neighs_{next} \leftarrow$ COMPUTENEIGHBORS(P_{next})
> $c \leftarrow$ COMPUTECOMMONNESS(P_{next}, τ_r)
> **if** $c > T_s$ **then**
> > **if** ISDISCRIMINATING($\mathscr{P}_{next}$, c, res) **then**
> > > $res \leftarrow res \cup \mathscr{P}_{next}$
> >
> > **if** $neighs_{next} \neq \emptyset$ **then**
> > > PATTERNMINE(P_{next}, $neighs_{next}$, vE, τ_r, T_s)
> >
> **else**
> > **if** $neighs_{next} \neq \emptyset$ **and** CANRISE($\mathscr{P}_{next}$) **then**
> > > PATTERNMINE($\mathscr{P}_{next}$, $neighs_{next}$, vE, τ_r, T_s)

---

are common (the *commonness* is high) in one subpopulation and rare (the *commonness* is low) in the other one.

As a second step, edges are sorted according to their average strength over the set of *SR-networks*, in order to find first patterns more representative and then potentially more interesting. Moreover, such a sorting is likely to make more effective succeeding pruning rules. Note that such a sorting changes during the analysis of the search space since for each pattern under consideration, the set of networks in which the pattern is relevant changes.

Next, the function PATTERNMINE, which is the core of the algorithm, is called. It receives the current pattern $\mathscr{P}_{cur}$, the neighbors of this patterns (which is the set of edges (v_i, v_j) such that either v_i or v_j are in $\mathscr{P}_{cur}$), and the set of already visited edges. The function tries to extend the current pattern by adding one neighbor at a time and computes its commonness. If its commonness is above the threshold, it checks if the pattern is discriminating according to Definition 4.5 and, in such a case, adds the pattern to the current result set.

Conversely, if the pattern commonness is below the threshold, since the measure is not monotone, it is still possible to find a discriminating pattern among its super-patterns. However, the function CANRISE can evaluate if such a super-pattern can exists. Intuitively, such a function considers the best edges among the remaining ones, and checks if the pattern obtained by adding these edges to the current pattern is interesting. The pattern built in this way is not guaranteed to exist since (i) the best edges could be not connected to the current pattern, and (ii) their strength is computed in the set of networks selected by the current pattern.

4.4.2 Strength Computation Details

We are interested in evaluating the maxima of f. Since the logarithmic function is a continuous monotone increasing one, and since adding a constant to a function does not alter the argument of the maximum, the maxima of f corresponds to the maximum of $g(x, y, \rho) = \log(f(x, y, \rho)) + \log(2\pi)$.
From

$$\log(f(x, y, \rho)) =$$

$$= \log\left(\frac{1}{2\pi\sqrt{1-\rho^2}}\right) + \left(-\frac{1}{2(1-\rho^2)}(x^2 + y^2 - 2\rho xy)\right) =$$

$$= -\log(2\pi) - \frac{1}{2}\log\left(1 - \rho^2\right) - \frac{1}{2}\left(\frac{x^2 + y^2 - 2\rho xy}{1 - \rho^2}\right)$$

we have that

$$g(x, y, \rho) = -\frac{1}{2}\log\left(1 - \rho^2\right) - \frac{1}{2}\left(\frac{x^2 + y^2 - 2\rho xy}{1 - \rho^2}\right).$$

Since we aim at finding the maxima of g as a function of ρ, we consider the partial derivative of g w.r.t. ρ

$$\frac{\partial g}{\partial \rho} = \frac{\rho}{1 - \rho^2} + \frac{(xy)(1 - \rho^2) - (x^2 + y^2 - 2\rho xy)\rho}{(1 - \rho^2)^2}$$

and we look for the stationary points by calculating the values of ρ where $\frac{\partial g}{\partial \rho} = 0$:

$$\frac{\rho}{1 - \rho^2} + \frac{(xy)(1 - \rho^2) - (x^2 + y^2 - 2\rho xy)\rho}{(1 - \rho^2)^2} = 0 \Longrightarrow$$

$$\rho^3 - \rho^2 xy + \rho(x^2 + y^2 - 1) - xy = 0. \qquad (4.2)$$

Therefore, we obtain a cubic equation. By setting

$$p = x^2 + y^2 - 1 - \frac{x^2 y^2}{3}$$

$$q = -xy + \frac{xy(x^2 + y^2 - 1)}{3} - \frac{2x^3 y^3}{27}$$

$$\Delta = \frac{q^2}{4} + \frac{p^3}{27},$$

we obtain that the solutions of (4.2) depend on the sign of Δ. In particular, the following two cases are possible:

$\Delta \geqslant 0$: We have just one real solution that, as can be easily verified, corresponds to a maximum:

$$\rho = \frac{xy}{3} + \sqrt[3]{-\frac{q}{2} + \sqrt{\Delta}} + \sqrt[3]{-\frac{q}{2} - \sqrt{\Delta}}$$

$\Delta < 0$: We should compute the square root of a negative number. This task has a solution in the set of complex numbers. Let define $z_1 = -\frac{q}{2} + i\sqrt{-\Delta}$ and $z_2 = -\frac{q}{2} - i\sqrt{-\Delta}$. Note that $z_1, z_2 \in \mathbb{C}$ and $z_2 = \overline{z_1}$. It follows that the solution of (4.2) can be written as

$$\rho = \frac{xy}{3} + \sqrt[3]{z_1} + \sqrt[3]{z_2}.$$

As there are three complex roots, there are three values for rho that are solution of (4.2)

$$\rho_k = \frac{xy}{3} + \sqrt[3]{|z_1| e^{i\frac{\vartheta + 2k\pi}{3}}} + \sqrt[3]{|z_2| e^{i\frac{-\vartheta + 2k\pi}{3}}}$$

with $k = 0, 1, 2$.
It follows that there are three solutions in $\mathbb{R}$

$$\rho_1 = \frac{xy}{3} \cos\left(\frac{\vartheta}{3}\right),$$

$$\rho_2 = \frac{xy}{3} \cos\left(\frac{\vartheta + 2\pi}{3}\right),$$

$$\rho_3 = \frac{xy}{3} \cos\left(\frac{\vartheta + 4\pi}{3}\right).$$

However, not all the solutions we have found are valid stationary points for g, because they are not in the function's domain. Therefore, among the valid values of ρ (i.e., $-1 \leq \rho \leq 1$), we have to choose only the one that maximizes g.

4.4.3 Relevance Computation Details

Consider Eq. (4.2) again. Given values ρ_0 and y_0, with $0 < \rho_0 \leq 1$, we aim at finding the values of x such that the value of ρ solution of Eq. (4.2) is larger than ρ_0.[3]

---

[3]Note that, due to the symmetry of Eq. (4.2), the same line of reasoning can be followed to find, given values ρ_0 and x_0, the values of y such that the value of ρ solution of Eq. (4.2) is larger than ρ_0.

Theorem 4.1 *Let ρ_0, x_0, and y_0 with $0 < \rho_0 \leq 1$ be such that Eq. (4.2) holds with this input. Let, also, x' and x'' be the solutions of Eq. (4.2), solved w.r.t. x, by setting $\rho = \rho_0$ and $y = y_0$. For any $x' \leq x \leq x''$, the value of ρ such that Eq. (4.2) holds is greater than ρ_0.*

Proof First of all note that Eq. (4.2) is a quadratic equation w.r.t. to x and then it cannot admit more than two solutions. Since ρ_0 is the solution of the equation when $y = y_0$, Eq. (4.2) admits for sure real solutions x' and x''.

Consider the function $\Psi(\rho, x)$ obtained from Eq. (4.2) by setting $y = y_0$. It implicitly defines a function $\rho = \psi(x)$ and, by construction, $\phi(x') = \phi(x'') = \rho_0$.

In order to prove the theorem, we prove that ψ is concave for any x between x' and x'' and, then, the value of ρ is greater than ρ_0.

Consider the first derivative of ψ w.r.t. x. Since ψ is implicitly defined, according to Dini's Theorem,

$$\frac{d\psi}{dx} = -\frac{\partial\Psi/\partial x}{\partial\Psi/\partial\rho} = -\frac{-\rho^2 + 2\rho x - y_0}{3\rho^2 - 2\rho x y_0 + (x^2 + y_0^2 - 1)}.$$

In order to study the growth of the function, we have to solve the following system:

$$-\begin{cases} \dfrac{-\rho^2 y_0 + 2\rho x - y_0}{3\rho^2 - 2\rho x y_0 + (x^2 + y_0^2 - 1)} \geq 0 & (4.3) \\ \rho^3 - \rho^2 x y_0 + \rho(x^2 + y_0^2 - 1) - x y_0 = 0. & (4.4) \end{cases}$$

First of all, note that the denominator of Eq. (4.3) is always greater than 0. Indeed, if $3\rho^2 - 2\rho x y_0 + (x^2 + y_0^2 - 1) < 0$ then $x^2 + y_0^2 - 1 = 2\rho x y_0 - 3\rho^2 - \varepsilon$ for some $\varepsilon > 0$. But this value makes unsatisfiable Eq. (4.4), since

$$\rho^3 - \rho^2 x y_0 + \rho(2\rho x y_0 - 3\rho^2 - \varepsilon) - x y_0 =$$
$$= (-2\rho^3 - x y_0(1 - \rho^2) - \varepsilon) < 0$$

for any $\varepsilon > 0$ and $\rho \leq 1$.

Thus, by considering just numerator, Eq. (4.3) is satisfied for any

$$x \leq \frac{(\rho^2 + 1)y_0}{2\rho} \tag{4.5}$$

and the derivative is 0 for $x = \bar{x} = \frac{(\rho^2+1)y_0}{2\rho}$.

By solving Eq. (4.4) w.r.t. x we obtain solutions

$$x' = \frac{1}{2\rho}\left(y_0(\rho^2+1) - \sqrt{y_o^2(\rho^2-1)^2 - 4\rho^2(\rho^2-1)}\right)$$
$$x'' = \frac{1}{2\rho}\left(y_0(\rho^2+1) + \sqrt{y_o^2(\rho^2-1)^2 - 4\rho^2(\rho^2-1)}\right)$$

(4.6)

By coupling Eq. (4.5) with Eq. (4.6), we obtain that x' is smaller than $\overline{x}$, x'' is larger than $\overline{x}$, and the function increases before $\overline{x}$ and decreases after $\overline{x}$. Thus, x is a maximum and all the values of x in $[x', x'']$ are such that $\psi(x) > \rho_0$. □

References

1. Allison, D.B., Cui, X., Page, G.P., Sabripour, M.: Microarray data analysis: from disarray to consolidation and consensus. Nat. Rev. Genet. **7**(1), 55–65 (2006)
2. Anastassiou, D.: Computational analysis of the synergy among multiple interacting genes. Mol. Syst. Biol. **3**(1), 83 (2007)
3. Atias, N., Sharan, R.: Comparative analysis of protein networks: hard problems, practical solutions. Commun. ACM **55**(5), 88–97 (2012)
4. Dehmer, M., Emmert-Streib, F., Graber, A., Salvador, A.: Applied statistics for network biology: methods in systems biology. John Wiley & Sons (2011)
5. Emmert-Streib, F., Tripathi, S., de Matos Simoes, R.: Harnessing the complexity of gene expression data from cancer: from single gene to structural pathway methods. Biol. Direct **7**(44.10), 1186 (2012)
6. Gray, R.M.: Entropy and information theory. Springer Science & Business Media (2011)
7. Metzker, M.L.: Sequencing technologies-the next generation. Nat. Rev. Genet. **11**(1), 31–46 (2010)
8. Mitchell, T.M.: Machine Learning, vol. 45. Burr Ridge, IL: McGraw Hill (1997)
9. Panni, S., Rombo, S.E.: Searching for repetitions in biological networks: methods, resources and tools. Brief. Bioinform. **16**(1), 118–136 (2015)
10. Quackenbush, J.: Computational analysis of microarray data. Nat. Revi. Genet. **2**(6), 418–427 (2001)
11. Roy, S., Bhattacharyya, D.K., Kalita, J.K.: Reconstruction of gene co-expression network from microarray data using local expression patterns. BMC Bioinform. **15**(Suppl 7), S10 (2014)
12. Rung, J., Brazma, A.: Reuse of public genome-wide gene expression data. Nat. Rev. Genet. **14**, 89–99 (2013)
13. Vidal, M., Cusick, M.E., Barabasi, A.L.: Interactome networks and human disease. Cell **144**(6), 986–998 (2011)
14. Wang, Z., Gerstein, M., Snyder, M.: Rna-seq: a revolutionary tool for transcriptomics. Nat. Rev. Genet. **10**(1), 57–63 (2009)
15. Watkinson, J., Wang, X., Zheng, T., Anastassiou, D.: Identification of gene interactions associated with disease from gene expression data using synergy networks. BMC Syst. Biol. **2**(1), 10 (2008)
16. Yan, X., Cheng, H., Han, J., Yu, P.S.: Mining significant graph patterns by leap search. In: ACM SIGMOD International Conference on Management of data, pp. 433–444. ACM (2008)